Green Energy and Technology

Hermann-Josef Wagner · Jyotirmay Mathur

Introduction to Wind Energy Systems

Basics, Technology and Operation

With 33 Figures and 6 Tables

 Springer

Prof.Dr. Hermann-Josef Wagner
Ruhr-Universität Bochum
Lehrstuhl für Energiesysteme
und Energiewirtschaft
Universtitätsstr. 150
44801 Bochum
Germany
lee@lee.rub.de

Dr.-Ing. Jyotirmay Mathur
Malaviya National Institute of
Technology
Dept. Mechanical Engineering
Jaipur-302017
JLN Marg
India
jyotirmay.mathur@gmail.com

ISBN 978-3-642-26047-6   e-ISBN 978-3-642-02023-0
DOI 10.1007/978-3-642-02023-0
Springer Heidelberg Dordrecht London New York

Springer Series in Green Energy and Technology ISSN 1865-3529

Cover design: WMXDesign GmbH, Heidelberg

Printed on acid-free paper

Springer is part of Springer Science+Business Media (www.springer.com)

Preface

In the past few decades, growth in the wind energy sector has been most phenomenal among all the renewable sources of energy. Consensus exists almost worldwide that for ensuring a sustainable future, wind energy can definitely play an important role.

The present book has been written to satisfy the interest of readers on wind energy converters. The authors have tried to strike a balance between a short book chapter and a very detailed book for subject experts. There were three prime reasons behind doing so: first, the field is quite inter-disciplinary and requires a simplified presentation for a person from non-parent discipline. The second reason for this short-version of a full book is that both authors have seen students and technically oriented people who were searching for this type of book on wind energy. The third reason and motivation for writing this book was to provide some initial information to people who are embarking on a career in the wind industry. It is this group of people that the present book is targeted at

This book presents the basic concepts of wind energy in its first two chapters. Chapter 3 deals with the physics and mechanics of operation. It describes the conversion of wind energy into rotation of turbine, and the critical parameters governing the efficiency of this conversion. Chapter 4 presents an overview of various parts and components of windmills with a blend of basics and recent advancements. Chapter 5 is dedicated to design considerations while selecting appropriate wind turbines for any site. Design options have been presented with their advantages and disadvantages. Chapter 6 is devoted to the utilization and control of operation of wind turbines. In this chapter, various parameters and methods for optimizing the performance are discussed. Chapter 7 presents the economics and financial issues associated with wind energy systems on the example of India and Germany.

Attempts have been made to include in this book technological advancements up to the beginning of the year 2009.

The authors thank Mrs. G. Schultz-Herzberger for controlling the English language of the manuscript. A special thank is given to Mrs. M. Koetter for her great help by typing and by controlling the text.

Authors also thank Prof. N.K. Bansal and Prof. R. Hanitsch for their suggestions. Support from DAAD for development of this book is also thankfully acknowledged.

The authors wish the readers a happy journey through the interesting field of wind energy.

Contents

List of Figures

List of Tables

About the Authors

Prof. Dr.-Ing. Hermann-Josef Wagner is postgraduate and doctorate in electrical and mechanical engineering from Technical University of Aachen, Germany. He is a Professor for Energy Systems and Energy Economics at the Ruhr-University of Bochum, Germany. He worked as a scientist for the Research Centre Juelich, for the German Parliament and for different universities. His relevant experiences are on the fields on energy systems analysis, renewable energies like wind energy and life cycle analysis, e-mail: lee@lee.ruhr-uni-bochum.de

Dr.-Ing. Jyotirmay Mathur is a mechanical engineer postgraduate in energy studies from Indian Institute of Technology, Delhi, India; and doctorate from University of Essen, Germany. He specializes in the areas of renewable energy systems, energy policy modelling and energy efficiency. Working as Reader in the Malaviya National Institute of Technology in Jaipur, Dr. Mathur has been founder coordinator of the postgraduate program in energy engineering at his institute. He is involved with several committees of national importance in India, e-mail: jyotirmay.mathur@gmail.com; jyotirmay@mnit.ac.in

Chapter 1
Wind Energy Today

1.1 Status

Wind energy is one of the oldest sources of energy used by mankind, comparable only to the use of animal force and biomass. Ancient cultures, dating back several thousand years, took advantage of wind energy to propel their sailing vessels. There are references to windmills relating to a Persian millwright in 644 AD and to windmills in Persia in 915 AD. These early wind energy converters were essential for pumping water and grinding cereals. In Europe, wind wheels were introduced around 1200 BC probably as an after-effect of the crusade to the orient. These European windmills were mainly used for grinding, except in the Netherlands where wind wheels supplied the power to pump river water to the land located below sea level. Between 1700 and 1800 AD the art of windmill construction reached its peak. The construction knowledge was relatively high and improved through trial and error. Later on, theories were developed, e.g. those by Euler, providing the tools to introduce new designs and, thus, to substantially improve the efficiency of energy conversion. Many windmills were built and operated in Denmark, England, Germany and the Netherlands during the eighteenth century. In 1750, the Netherlands alone had between 6,000 and 8,000 windmills in operation. The number of windmills in Germany has been estimated at about 18,000 in 1895, 11,400 in 1914, and between 4,000 and 5,000 in 1933.

Around the beginning of the twentieth century, windmills were further improved and the design of a multi-blade farm windmill originated in the USA. By the middle of the century, more than 6 million windmills were in operation in the USA. Worldwide, many of these wind wheels were used to produce mechanical power or as decentralized electricity suppliers on large farms. When the central electricity grid reached every farmhouse at the beginning of the twentieth century, the use of electricity produced by windmills rapidly decreased and the converters were taken out of operation as soon as the next repair job was due. In the nineteen fifties, pioneers like Huetter at the University of Stuttgart, Germany, took up developing and testing modern wind wheels again. Their design is quite different from the previous one; there are only two or three blades with very good aerodynamic parameters, able to rotate at high speed. Due to the high number of rotations, only a small generator is needed to produce electricity. Nevertheless, it was not possible to break

H.-J. Wagner, J. Mathur, *Introduction to Wind Energy Systems*, Green Energy and Technology, DOI 10.1007/978-3-642-02023-0_1, © Springer-Verlag Berlin Heidelberg 2009

Table 1.1 Use of wind energy world wide

Status of installed wind power	Rated capacity 1.1.2009 (MW)	Share worldwide (%)
USA	25,200	21
Germany	23,900	20
Spain	16,800	14
China	12,200	10
India	9,600	8
Italy	3,700	3
France	3,400	3
UK	3,200	3
Denmark	3,100	2
Portugal	2,800	2
Remaining countries	16,700	14
Total	120,600	100

even economically selling wind generated electricity in the fifties and sixties of the twentieth century. In the aftermath of the so called oil crisis in the seventies, there was a surge to enforce the development and marketing of wind wheels, especially in the USA, Denmark and Germany. This was based on the understanding that ultimately, additional energy sources emitting less pollution would be necessary. Due to favorable tax regulations in the eighties, about 12,000 wind converters supplying power ranging from 20 kW to about 200 kW were installed in California. In Europe, a lot of tax money was spent on the development of bigger wind converters and on marketing them. Now, in 2009, worldwide more than 100,000 windmills of about 120 GW installed power generation capacity are in operation, as shown in Table 1.1.

Germany had the leadership until 2007. Then USA has taken over the leadership. On the market are converters with a capacity up to 6 MW.

The market introduction of wind energy is proceeding in industrialized countries as well as in development countries like e.g. India. The Indian wind energy sector had an installed capacity of 9,600 MW as end of 2008. In terms of wind power installed capacity, India is ranked 5th in the World. Today, India is a major player in the global wind energy market. Still, the total potential of wind energy in the Indian country is far from exhausted. With the current level of technology, the "onshore" potential for utilization of wind energy for electricity generation is estimated of about 48,000 MW.

Up to the year 2008, the top five countries in terms of installed capacity are the USA, Germany, Spain, China, and India.

1.2 Advantages and Disadvantages of Wind Energy Systems

Wind energy offers many advantages, which explains why it is the fastest-growing energy source in the world. Research efforts are aimed at addressing the challenges to increase the use of wind energy.

1.2.1 Advantages

- Wind energy systems are energized by the naturally flowing wind, therefore it can be considered as a clean source of energy. Wind energy does not pollute the air like power plants that rely on combustion of fossil fuels, such as coal or natural gas. Wind turbines do not produce atmospheric emissions that cause acid rain or greenhouse gases.
- Wind energy is available as a domestic source of energy in many countries worldwide and not confined to only few countries, as in case of oil.
- Wind energy is one of the lowest-priced renewable energy technologies available today.
- Wind turbines can also be built on farms or ranches, thus benefiting the economy in rural areas, where most of the best wind sites are found. Farmers and ranchers can continue to use their land because the wind turbines use only a small fraction of the land. Wind power plant owners make rent payments to the farmer or rancher for the use of the land.

1.2.2 Disadvantages

- Wind power has to compete with conventional power generation sources on a cost basis. Depending on the wind profile at the site, the wind farm may or may not be as cost competitive as a fossil fuel based power plant. Even though the cost of wind power has decreased in the past 10 years, the technology requires a higher initial investment than fossil-fueled solutions for power supply.
- The major challenge to using wind as a source of power is that the wind is intermittent and it does not always blow when electricity is needed. Wind energy cannot be stored; and not all winds can be harnessed to meet the timing of electricity demands. The option of energy storage in battery banks is much beyond economically feasible limits for large wind turbines.
- Good wind sites are often located in remote locations, far from cities where the electricity is needed. In developing countries, there is always the extra cost of laying grid for connecting remote wind farms to the supply network.
- Wind resource development may compete with other uses for the land and those alternative uses may be more highly valued than electricity generation.
- Although wind power plants have relatively little impact on the environment compared to other conventional power plants, there is some concern over the noise produced by the rotor blades, and aesthetic (visual) impacts. Most of these problems have been resolved or greatly reduced through technological development or by properly siting wind plants.

1.3 Different Types of Wind Energy Converters: An Overview

Today, various types of wind energy converters are in operation (Fig. 1.1 gives an overview). The most common device is the horizontal axis converter. This converter consists of only a few aerodynamically optimized rotor blades, which for the

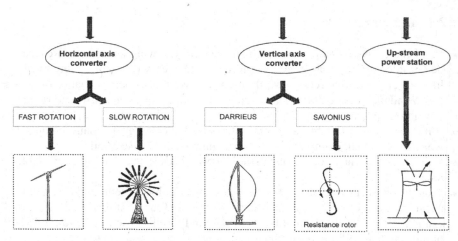

Fig. 1.1 Overview of different types of wind energy converters

purpose of regulation can usually be turned about their long axis (Pitch-regulation). Another cheaper way of regulation consists in designing the blades in such a way that the air streaming along the blades will go into turbulence at a certain speed (Stall-Regulation). These converters can deliver power ranging from 10 kW to some MW. The largest converter on the European market has a power of 6 MW. The efficiency of this type of wind energy converters in comparison with other types of windmills is very high. Another conventional (older) type of horizontal axis rotor is the multi-blade wind energy converter. It was first built about 100 years ago. Such windmills have a high starting torque which makes them suitable for driving mechanical water pumps. The number of rotations is low, and the blades are made from simple sheets with an easy geometry. For pumping water, a rotation regulating system is not necessary, but there is a mechanical safety system installed to protect the converter against storm damage. The rotor is turned in the direction of the wind by using a so called wind-sheet in leeward direction. The mechanical stability of such "slow speed converters" is very high; some have had operation periods of more than 50 years. In order to increase the number of rotations, this type of converter had been improved and equipped with aerodynamically more efficient blades facilitating the production of electricity, where the area of a blade is smaller.

A third type of converter is known as DARRIEUS, a vertical axis construction. Their advantage is that they do not depend on the direction of the wind. To start, they need because of their low starting torque the help of a generator working as a motor or the help of a SAVONIUS rotor installed on top of the vertical axis. In the 1980s, a reasonable number of Darrieus-converters were installed in California, but a further expansion into the higher power range and worldwide has not taken place. One reason may be that wind velocity increases significantly with height, making horizontal axis wheels on towers more economical. A modification of the Darrieus rotor is in the form of H-rotor; there are more than 30 installations of H-rotors worldwide but all of them are below the capacity of 300 kW. Prototype testing was successfully completed for this type but the commercial stage is yet to be seen. The SAVONIUS rotor is used as a measurement device especially for wind velocity; it is used for power production for very small capacities under 100 W.

The last technique to deal with is known as Up-Stream-Power-Station or thermal tower. In principle, it can be regarded as a mix between a wind converter and a solar collector. The top of a narrow, high tower contains a wind wheel on a vertical axis driven by the rising warm air. A solar collector installed around the foot of the tower heats up the air. The design of the collector is simple; a transparent plastic foil is fixed several meters above the ground in a circle around the tower. Therefore, the station needs a lot of space and the tower has to be very high. Such a system has a very poor efficiency, only about 1%. The advantage of such a design is its technical simplicity, which may enable developing countries to construct it by themselves. Worldwide, only one Up-Stream-Power-Station, designed by a German company, has been built so far. For some years it worked satisfactorily at the location of Manzanares in Spain, but in the mid eighties it was destroyed by bad weather. This station had an electrical power of 50 kW, the tower was about 200 m high and the collector had a diameter of approximately the same size. Another Up-Stream-Power-Station with an electrical performance of about 200 MW was planned in India. Other feasibility studies have also been conducted in Australia and Namibia but no project of this type has been realized.

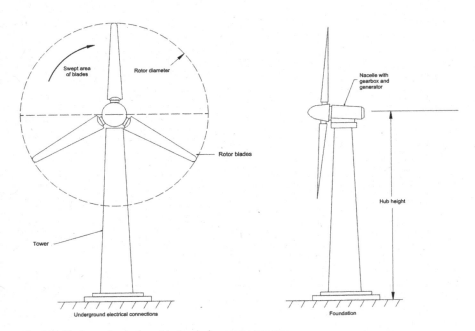

Fig. 1.2 Horizontal axis three blade wind energy converter

Throughout this book, the three terms "wind energy converters", "windmills" and "wind turbines" have been used quite interchangeably. The first term is the technical name of the system, whereas the other two are popularly used terms. Figure 1.2 shows the most popular type of horizontal axis three blade wind energy converter for generating electricity, worldwide. This figure shows the front and side views of a three blade horizontal axis wind energy converter. Details about its major parts and their working are discussed in subsequent chapters.

Chapter 2
Wind: Origin and Local Effects

All renewable energy (except tidal and geothermal power), and even the energy in fossil fuels, ultimately comes from the sun. About 1–2% of the energy coming from the sun is converted into wind energy. This chapter explains the cause of wind flow and factors that affect the flow pattern. Understanding these is necessary for selecting proper locations for wind turbines.

2.1 Origin and Global Availability

The regions around the equator, at 0° latitude, are heated more by the sun than the rest of the globe. Hot air is lighter than cold air; it will rise into the sky until it reaches approximately 10 km altitude and will spread to the North and the South. If the globe did not rotate, the air would simply arrive at the North Pole and the South Pole, sink down, and return to the equator.

Since the globe is rotating, any movement on the Northern hemisphere is diverted to the right, if we look at it from our own position on the ground. (In the Southern hemisphere it is bent to the left). This apparent bending force is known as the Coriolis force, named after the French mathematician Gustave Gaspard Coriolis.

In the Northern hemisphere, the wind tends to rotate counterclockwise as it approaches a low pressure area. In the Southern hemisphere, the wind rotates clockwise around low pressure areas.

The wind rises from the equator and moves north and south in the higher layers of the atmosphere. Around 30° latitude in both hemispheres the Coriolis force prevents the air from moving much farther. At this latitude, there is a high pressure area, as the air begins to sink down again. As the wind rises from the equator there is a low pressure area close to ground level attracting winds from the North and South. At the Poles, there is high pressure due to the cooling of the air. Table 2.1 shows the prevailing direction of global winds.

The prevailing wind directions are important when siting wind turbines, since one would obviously want to place them in areas with the least obstacles from the prevailing wind directions.

H.-J. Wagner, J. Mathur, *Introduction to Wind Energy Systems*, Green Energy and Technology, DOI 10.1007/978-3-642-02023-0_2, © Springer-Verlag Berlin Heidelberg 2009

Table 2.1 Prevailing global wind directions

Latitude	90–60°N	60–30°N	30–0°N	0–30°S	30–60°S	60–90°S
Direction	NE	SW	NE	SE	NW	SE

2.2 Local Effects on Wind Flow

Winds are very much influenced by the ground surface at altitudes up to 100 m. The winds are slowed down by the earth's surface roughness and obstacles. There may be significant differences between the direction of the global or geostrophic winds because of the earth's rotation (the Coriolis force), and the wind directions near the surface. Close to the surface of the earth, the following effects influence the flow pattern of wind:

(a) *Sea breezes.* Land masses are heated by the sun more quickly than the sea in the daytime. The air rises, flows out to the sea, and creates a low pressure at ground level which attracts the cool air from the sea. This is called a sea breeze. At nightfall there is often a period of calm when land and sea temperatures are equal. At night the wind blows in the opposite direction. The land breeze at night generally has lower wind speeds, because the temperature difference between land and sea is smaller at night.

The monsoon (specific period of the year when majority of rainfall occurs) in India and all of South-East Asia is in reality a large-scale form of the sea breeze and land breeze, varying in its direction between seasons, because land masses are heated or cooled more quickly than the sea.

(b) *Mountain breeze.* Mountain regions display many interesting weather patterns. One example is the valley wind which originates on South-facing slopes (North-facing in the Southern hemisphere). When the slopes and the neighboring air are heated, the density of the air decreases, and the air ascends towards the top following the surface of the slope. At night the wind direction is reversed, and turns into a down-slope wind. If the valley floor is sloped, the air may move down or up the valley, as a canyon wind. Winds flowing down the leeward sides of mountains can be quite powerful.

(c) *The wind rose.* It can be noticed that strong winds usually come from a particular direction. To show the information about the distributions of wind speeds, and the frequency of the varying wind directions, one may draw a so-called wind rose as shown in Fig. 2.1 on the basis of meteorological observations of wind speeds and wind directions.

The wind rose presents a summary of annual wind data. The circular space can e.g. be divided in 16 sectors representing major directions from which wind might come. The number of segments may be more but it makes the diagram difficult to read and interpret. The concentric circles having percent values represent the probability of wind coming from any particular direction. In Fig. 2.1, the polygon in north direction goes slightly ahead of the 10% circle, hence it can be concluded that at this location, the probability of wind coming from the North is roughly 12%. The

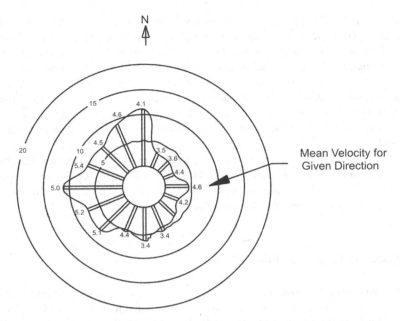

Fig. 2.1 Example of a wind rose (Annual all directions average velocity 4.3 m/s)

bars going up to the border of the polygon have a value tag with them. This value represents the mean velocity of wind when it comes from that particular direction. Again, in Fig. 2.1, the bar in the North direction says that when wind comes from the North it has an average velocity of 4.1 m/s. This graph shows that the strongest wind comes from the westward directions.

It should be noted that in popular terms, when we say that some site has "North wind", this means that wind is coming from the North and not going to the North.

Further to note is that wind patterns may vary from year to year, and the energy content may vary (typically by some 10%) from year to year, so it is best to have observations from several years to calculate a credible average. Planners of large wind parks will usually rely of local measurements, and use long-term meteorological observations from nearby weather stations to adjust their measurements to obtain a reliable long term average.

2.3 Attractive Locations for Wind Energy

In Europe, the area close to the North Sea has the strongest winds. In the extreme North, a wind velocity of the order of 10–11 m/s is available at a height of 50 m. This velocity is very attractive for installation wind machines. The Southern part of Europe also has wind energy potential but it has relatively less wind velocities, in the range of 3–8 m/s at the height of 50 m above the ground.

The Centre for Wind Energy Technology, India, has estimated a 48,000 MW potential of wind energy in India. This estimate is based on the current technologies and economics of wind energy. The Southern part of India is most attractive for wind energy. The wind energy map of India reveals that the maximum wind energy potential lies either in the Southern or Western part of India. In the states of Tamilnadu and Karnataka, there are sites having velocities in the range of 6–8 m/s at a 50 m height above the ground. But in the Western parts of India, close to the Arabian Sea as well as in the desert of Rajasthan, attractive wind energy locations also exist.

The wind data available from various agencies give a fairly good idea about attractive locations for wind energy installations. However, to get exact knowledge about certain locations, measurement of wind availability over several years would be appropriate. The reasons for this are explained in the next section.

2.4 Local Effects on Wind Flow

For the purpose of wind turbines, local wind is more important, since due to local effects, a site may have very low wind even if it is situated in a predominantly windy area. Major factors that govern local winds are, therefore, described in this section.

2.4.1 Roughness Length and Wind Shear

High above ground level the wind is influenced by the surface of the earth at all. In the lower layers of the atmosphere, however, wind speeds are affected by the friction against the surface of the earth. In the wind industry one distinguishes between the roughness of the terrain, the influence from obstacles, and the influence from the terrain contours, which is also called the orography of the area.

The more pronounced the roughness of the earth's surface, the more the wind will be slowed down. In the wind industry, wind conditions in a landscape are referred through roughness classes or roughness lengths. The term roughness length is the distance above ground level where the wind speed theoretically should be zero. A high roughness class of 3–4 refers to landscapes with many trees and buildings, while a sea surface is in roughness class 0. Concrete runways in airports are in roughness class 0.5.

2.4.2 Wind Speed Variability

The wind speed is always fluctuating, and thus the energy content of the wind is always changing. The variation depends on the weather and on local surface conditions and obstacles. Energy output from a wind turbine will vary as the wind varies,

Fig. 2.2 Short term
variability of the wind

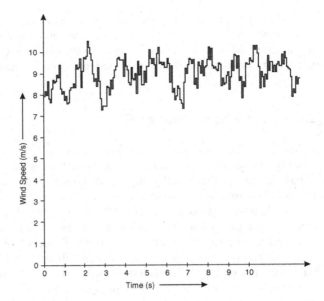

although the most rapid variations will to some extent be compensated for by the inertia of the wind turbine rotor. Figure 2.2 shows short term variations in wind.

In most locations around the globe it is more windy during the daytime than at night. This variation is largely due to the fact that temperature differences, e.g. between the sea surface and the land surface, tend to be larger during the day than at night. The wind is also more turbulent and tends to change direction more frequently during the day than at night.

From the point of view of wind turbine owners, it is an advantage that most of the wind energy is produced during the daytime, since electricity consumption is higher than at night. Many power companies pay more for the electricity produced during the peak load hours of the day (when there is a shortage of cheap generating capacity).

At most locations, the wind may not be sufficient for producing power continuously for two to three days and sometimes even for one week.

2.4.3 Turbulence

It is normally experienced that hailstorms or thunderstorms in particular, are associated with frequent gusts of wind which both change speed and direction. In areas with a very uneven terrain surface, and behind obstacles such as buildings, a lot of turbulence is similarly created, with very irregular wind flows, often in whirls or vortexes in the neighborhood.

Turbulence decreases the possibility of using the energy in the wind effectively for a wind turbine. It also imposes more tear and wear on the wind turbine, as explained in the section on fatigue loads. Towers for wind turbines are usually made tall enough to avoid turbulence from the wind close to ground level.

2.4.4 Obstacles to Wind Flow

Obstacles to the wind such as buildings, trees, rock formations etc. can decrease wind speeds significantly, and they often create turbulence in their neighborhood. It can be seen in Fig. 2.3 that in case of typical wind flows around an obstacle, the turbulent zone may extend to some three times the height of the obstacle. The turbulence is more pronounced behind the obstacle than in front of it. Therefore, it is best to avoid major obstacles close to wind turbines, particularly if they are upwind in the prevailing wind direction, i.e. "in front of" the turbine.

However, every tower of a wind converter will work as such an obstacles and will have influence on the blade when it is coming before or behind the tower.

Obstacles will decrease the wind speed downstream. The decrease in wind speed depends on the porosity of the obstacle, i.e. how "open" the obstacle is. Porosity is defined as the open area divided by the total area of the object facing the wind. A building is obviously solid, and has no porosity, whereas a fairly open tree may let more than half of the wind through. In case of very dense trees, the porosity is less, say one third. The slowdown effect on the wind from an obstacle increases with the height and length of the obstacle. The effect is obviously more pronounced close to the obstacle, and close to the ground.

(a) *Reduction in obstacles with turbine hub height.* The higher a turbine is above the top of the obstacle, the less wind shade will be produced. The wind shade, however, may extend to up to five times the height of the obstacle at a certain distance. If the obstacle is taller than half the hub height, the results are more uncertain, because the detailed geometry of the obstacle, (e.g. differing slopes of the roof on buildings) will affect the result.

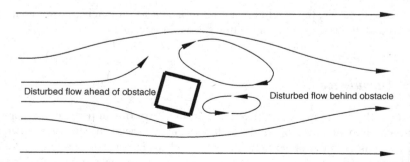

Fig. 2.3 Obstacle in wind flow

(b) *Distance between obstacle and turbine*. The distance between the obstacle and the turbine is very important for the shelter effect. In general, the shelter effect will decrease as one moves away from the obstacle, just like a smoke plume becomes diluted as we move away from a smokestack. In terrain with very low roughness (e.g. water surfaces), the effect of obstacles (e.g. an island) may be measurable up to 20 km away from the obstacle. If the turbine is closer to the obstacle than five times the obstacle height, the results will be more uncertain, because they will depend on the exact geometry of the obstacle.

2.4.5 The Wind Wake and Park Effect

Since a wind turbine generates electricity from the energy in the wind, the wind leaving the turbine must have a lower energy content than the wind arriving in front of the turbine. There will be a wake effect behind the turbine, i.e. a long trail of wind which is quite turbulent and slowed down, as compared to the wind arriving in front of the turbine. The expression wake is obviously derived from the wake behind a ship. Wind turbines in parks are usually spaced at least three rotor diameters from one another in order to avoid too much turbulence around the turbines downstream.

As a result of the wake effect, each wind turbine will slow down the wind behind it as it pulls energy out of the wind and converts it to electricity. Ideally, therefore turbines should be spaced as far apart as possible in the prevailing wind direction. On the other hand, land use and the cost of connecting wind turbines to the electrical grid would force to space them closer together.

As a guideline for wind park design turbines in wind parks are usually spaced somewhere between 5 and 9 rotor diameters apart in the prevailing wind direction, and between 3 and 5 diameters apart in the direction perpendicular to the prevailing winds. In Fig. 2.4, three rows of five turbines each are placed in a fairly typical pattern. The turbines (the dots) are placed 7 diameters apart in the prevailing wind direction, and 4 diameters apart in the direction perpendicular to the prevailing winds.

2.4.6 The Hill Effect and Tunnel Effect

(a) *Speed up effects: tunnel effect*. While walking between tall building, or in a narrow mountain pass, it can be noticed that wind velocity increases. The air becomes compressed on the windy side of the buildings or mountains, and its speed increases considerably between the obstacles to the wind. This is known as a "tunnel effect". So, even if the general wind speed in open terrain may be, say, 6 m/s, it can easily reach 9 m/s in a natural "tunnel". Placing a wind turbine in such a tunnel is one clever way of obtaining higher wind speeds than in the surrounding areas. To obtain a good tunnel effect the tunnel should be "softly" embedded in the landscape. In case the hills are very rough and uneven, there may be lots of turbulence in the area,

Fig. 2.4 Spacing between
turbines in a wind park in
terms of rotor diameters (e.g.
4 means four times the rotor
diameter)

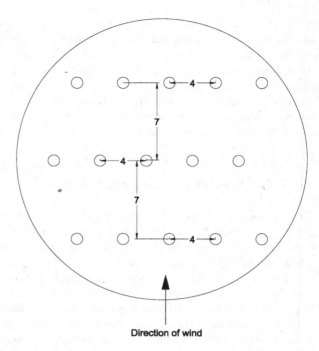

i.e. the wind will be whirling in a lot of different (and rapidly changing) directions. If there is much turbulence it may negate the wind speed advantage completely, and the changing winds may inflict a lot of useless tear and wear on the wind turbine.

(b) *The hill effect*. A common way of siting wind turbines is to place them on hills or ridges overlooking the surrounding landscape. In particular, it is always an advantage to have as wide a view as possible in the prevailing wind direction in the area. On hills, one may also experience that wind speeds are higher than in the surrounding area. Once again, this is due to the fact that the wind becomes compressed on the windy side of the hill, and once the air reaches the ridge it can expand again as its soars down into the low pressure area on the lee side of the hill.

2.5 Selecting a Turbine Site

Looking at nature itself is usually an excellent guide to finding a suitable wind turbine site. However, some typical considerations are as follows:

(a) *Wind conditions*. If there are trees and shrubs in the area, one may get a good clue about the prevailing wind direction. While moving along a rugged coast-line, centuries of erosion which have worked in one particular direction can also be noticed. Meteorology data, ideally in terms of a wind rose calculated over 20–25 years is probably the best guide, but these data are rarely collected directly at the site, and there are many reasons to be careful about the use of meteorology data. Meteorologists collect wind data for weather forecasts and aviation, and that information is often used to assess the general wind conditions for wind energy in an area. Unless calculations are made which compensate for the local conditions under which the meteorology measurements were made, it is difficult to estimate wind conditions at a nearby site. In most cases using meteorology data directly will underestimate the true wind energy potential in an area.

If there are already wind turbines in the area, their production results are an excellent guide to local wind conditions. In countries like Denmark, Germany, Spain, and in the Southern part of India, where a large number of turbines are found scattered around the countryside, manufacturers can offer guaranteed production results on the basis of wind calculations made on the site.

(b) *Look for a view.* It is often preferred to have a view as wide and open as possible in the prevailing wind direction, and we would like to have as few obstacles and as low a roughness as possible in that same direction. If a rounded hill can be found to place the turbines, one may even get a speed up effect in the bargain.

(c) *Soil conditions and transportation facilities.* Both the feasibility of building foundations for the turbines and road construction to reach the site with heavy trucks must be taken into account with any wind turbine project. Due to the large size of equipment and machinery, these sometimes become bottlenecks for the installation of systems.

To sum up the coverage on wind, it may be said that for taking a macro level decision such as around which city the installation should be planned, the global wind availability and published wind data may be used. However, for an exact siting of wind energy system, a micro level analysis would be necessary due to large variations caused by the local geographical details. Only in case of very high turbines, e.g. 130 m high, where local disturbances have nearly no effect, it may not be necessary to go into micro level details. The exercise of finding an exact location for installation is termed as "micro-siting" which requires actual measurement of wind data at the site.

Chapter 3
Physics of Wind Energy

The basic principles of physics on which any wind turbine works are explained in this chapter. These concepts will be helpful in understanding the science and technology behind the operation and control of wind turbines in order to harvest maximum energy from the wind.

3.1 Energy Content in Wind

A wind turbine obtains its power input by converting the force of the wind into torque (turning force) acting on the rotor blades. The amount of energy which the wind transfers to the rotor depends on the density of the air, the rotor area, and the wind speed.

(a) *Density of air*. The kinetic energy of a moving body is proportional to its mass. The kinetic energy in the wind thus depends on the density of the air, i.e. its mass per unit of volume.

In other words, the "heavier" the air, the more energy is received by the turbine. At normal atmospheric pressure and at 15°C, the density of air is 1.225 kg/m^3, which increases to 1.293 kg/m^3 at 0°C and decreases to 1.164 kg/m^3 at 30°C. In addition to its dependence upon temperature, the density decreases slightly with increasing humidity. At high altitudes (in mountains), the air pressure is lower, and the air is less dense. It will be shown later in this chapter that energy proportionally changes with a variation in density of air.

(b) *Rotor area*. When a farmer tells how much land he is farming, he will usually state an area in terms of square meters or hectares or acres. With a wind turbine it is much the same story, though wind farming is done in a vertical area instead of a horizontal one. The area of the disc covered by the rotor (and wind speeds, of course), determines how much energy can be harvested over a year.

A typical 1,000 kW wind turbine has a rotor diameter of 54 m, i.e. a rotor area of some 2,300 m^2. The rotor area determines how much energy a wind turbine is able to harvest from the wind. Since the rotor area increases with the square of the rotor diameter, a turbine which is twice as large will receive 2^2, i.e. four times as much energy.

H.-J. Wagner, J. Mathur, *Introduction to Wind Energy Systems*, Green Energy and Technology, DOI 10.1007/978-3-642-02023-0_3, © Springer-Verlag Berlin Heidelberg 2009

Fig. 3.1 Power output increases with the rotor diameter and the swept rotor area

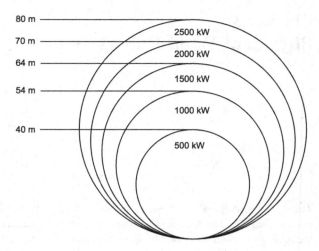

Figure 3.1 gives an idea of the normal rotor sizes of wind turbines: If the rotor diameter is doubled, one gets an area which is four times larger (two squared). This means that four times as much power output from the rotor will also be obtained.

Rotor diameters may vary somewhat from the figures given above, because many manufacturers optimize their machines to local wind conditions: A larger generator, of course, requires more power (i.e. strong winds) to turn at all. So if one installs a wind turbine in a low wind area, annual output will actually be maximized by using a fairly small generator for a given rotor size (or a larger rotor size for a given generator). The reason why more output is available from a relatively smaller generator in a low wind area is that the turbine will be running more hours during the year.

(c) *Wind velocity*. Considering an area A (e.g. swept area of blades) and applying a wind velocity v, the change in volume with respect to the length "l" is:

$$\Delta V = A \cdot \Delta l,$$
$$v = \frac{\Delta l}{\Delta t} \tag{3.1}$$
$$\Rightarrow \Delta V = A \cdot v \cdot \Delta t.$$

The energy in the wind is in the form of kinetic energy. Kinetic energy is characterized by the equation:

$$E = \frac{1}{2}mv^2 \tag{3.2}$$

The change in energy is proportional to the change in mass, where

$$m = V \cdot \rho_a \tag{3.3}$$

and ρ_a the specific density of the air. Therefore, substituting for V and m yields

$$E = \frac{1}{2} \cdot A \cdot \rho_a \cdot v^3 \cdot t. \tag{3.4}$$

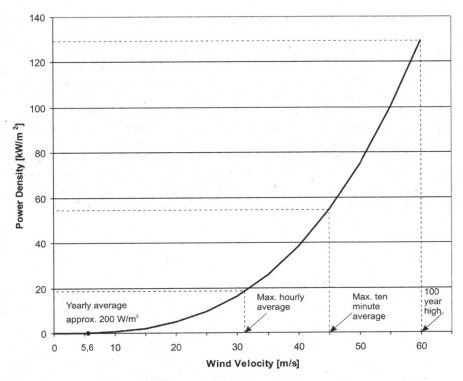

Fig. 3.2 Relationship between wind velocity and power of wind (wind speed for Germany)

From the previous equation it can be seen that the energy in the wind is proportional to the cube of the wind speed, v^3. The power P is defined as

$$P = \frac{E}{t} = \frac{1}{2} \cdot A \cdot \rho_a \cdot v^3 \qquad (3.5)$$

Therefore, power in wind is proportional to v^3. From Fig. 3.2 it can be seen that the power output per m^2 of the rotor blade is not linearly proportional to the wind velocity, as proven in the theory above. This means that it is more profitable to place a wind energy converter in a location with occasional high winds than in a location where there is a constant low wind speed. Measurements at different places show that the distribution of wind velocity over the year can be approximated by a Weibull-equation. This means that at least about 2/3 of the produced electricity will be earned by the upper third of wind velocity. From a mechanical point of view, the power density range increases by one thousand for a variation of wind speed of factor 10, thus producing a construction limit problem. Therefore, wind energy converters are constructed to harness the power from wind speeds in the upper regions.

3.2 Energy Conversion at the Blade

Figure 3.3 shows the velocities and forces at the profile of a rotating blade. The blade itself moves with an average circumferential velocity u in the plane of the rotor. Rotation of blade provides a relative velocity of air with respect to the blade, which can be considered as additional wind (velocity vector u) working together with the actual wind to determine the rotational force on the blade. The wind flows perpendicular to the plane of the rotor, thus creating a resultant velocity vector w. This velocity is the relative approach or flow velocity of the rotor blade.

The two main forces acting on the rotor blade are the lift force F_A and the drag force F_R. The drag force acts parallel to the initial direction of movement and the lift force acts perpendicular to it. The lift force is the greater force in normal operating conditions and arises due to the unequal pressure distribution around an aerofoil

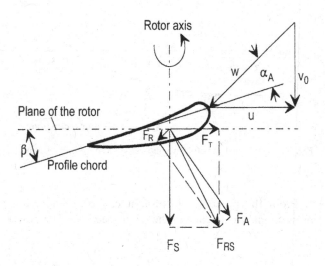

α_A = Angle of attack

β = Pitch Angle

u = Average circumferential velocity

v_0 = Wind velocity in the rotor plane

w = Relative approach velocity

F_R = Drag force

F_A = Lift force

F_{RS} = Resultant force

F_T = Tangential component

F_S = Axial component

Fig. 3.3 The velocities and forces acting on a blade

profile. The pressure on the upper surface is lower than that on the underside, therefore the air has a higher velocity when passing over the upper surface of the profile. The lift force is determined by the following formula:

$$F_A = \frac{1}{2} \cdot \rho_a \cdot c_A \cdot w^2 \cdot A, \tag{3.6}$$

where c_A is the lift force coefficient. The drag force is determined from a similar formula,

$$F_R = \frac{1}{2} \cdot \rho_a \cdot c_R \cdot w^2 \cdot A, \tag{3.7}$$

where c_R is the drag force coefficient, and is caused by air friction at the surface of the profile. The relationship between the two forces is given by the ratio E_G of their coefficients,

$$E_G = \frac{c_A}{c_R}. \tag{3.8}$$

This relation is at least given by the aerodynamic quality of the blade, by its design and its surface quality. It can be seen from Fig. 3.3 that the resultant force F_{RS} of the lift and drag forces can be divided into two components: the tangentially acting component F_T and the axially acting component F_S. It is the force F_T that causes the rotation of the rotor blade and makes power delivery possible.

3.3 Power Coefficients and Principles of Design

The wind turbine rotor must obviously slow down the wind as it captures its kinetic energy and converts it into rotational energy. This means that the wind will be moving slowly after leaving the rotor as compared to its movement before entering the rotor. Farther downstream, the turbulence in the wind will cause the slow wind behind the rotor to mix with the faster moving wind from the surrounding area. The wind shade behind the rotor will therefore gradually diminish as one moves away from the turbine. The factors governing the proportion of velocity ahead and behind the rotor that in turn govern the power output, are discussed in the following sub-sections:

3.3.1 Coefficient of Power c_p and Betz' Law

The question of how much of the wind energy can be transferred to the blade as mechanical energy has been answered by the Betz' law.

Betz' law states that only a maximum of 59.25% of the kinetic power in the wind can be converted to mechanical power using a wind turbine, the so called maximum power coefficient or Betz-Number. This number is not higher because the wind on the back side of the rotor must have a high enough velocity to move away and allow more wind through the plane of the rotor.

The relationship between the mechanical power of the rotor blade P_R and the power of wind P in the rotor area is given by the power coefficient c_p:

$$c_p = \frac{P_R}{P} \tag{3.9}$$

The power coefficient c_p can be interpreted as the efficiency between the rotor blade and the wind. The maximum power coefficient, the above mentioned Betz number, determined by the ratio $v_2/v_1 = 1/3$. Therefore, an ideal turbine will slow down the wind by 2/3 of its original speed.

3.3.2 Tip Speed Ratio

A wind energy converter is classified through the characteristic tip speed ratio λ_S. This is the ratio (as a scalar) of the circumferential velocity of the rotor at the end of the blade (maximum velocity u_e) and the wind velocity v_0 in front of the rotor blade:

$$\lambda_S = \frac{u_e}{v_0}. \tag{3.10}$$

The tip speed ratio has a strong influence on the efficiency of a wind energy converter (Fig. 3.4). When λ_S is small, the circumferential velocity is also small which results in an increase in the angle of attack α_A. When the angle of attack increases past a critical angle, the wind flow breaks away from the blade profile and becomes turbulent, thus dramatically reducing the lift force. If the tip speed ratio is too large, the lift force will reach its maximum value and decrease afterwards, thus reducing the power efficiency of the converter.

3.3.3 Power Efficiency

The power efficiency of a rotor blade can be determined by calculating the relationship between the power coefficient and the tip speed ratio. Figure 3.5 shows that for every pitch angle β, there is a tip speed ratio λ_S which corresponds to the maximum power coefficient and hence the maximum efficiency. It can be seen that the power efficiency significantly depends on the pitch angle and the tip speed ratio. Therefore, the pitch angle of the blade has to be changed mechanically in respect to the actual tip speed ratio. In case of any pitch angles the power coefficient is negative. These means that the rotor will not turn. It works as a brake.

A disk brake on the main shaft is therefore not necessary in most modern converters by using pitch control.

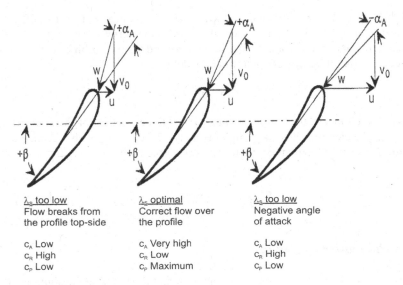

Fig. 3.4   Influence of tip-speed ratio λ_S

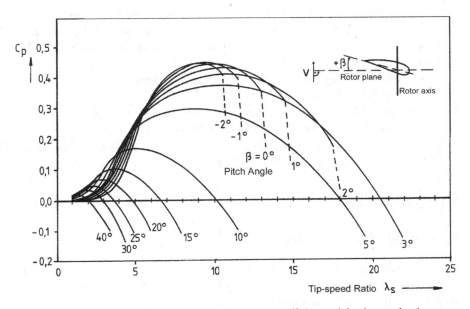

Fig. 3.5   Example of the relationship between the power coefficient and the tip-speed ratio

Besides the power coefficient c_p which can be interpreted as the efficiency between the rotor blade and the wind, there are also energy losses in the mechanical components of the rotor and gear system and in the turbine and generator connection. Therefore, the efficiency can be defined as:

$$\eta = c_p \cdot \eta_m \cdot \eta_{ge}, \tag{3.11}$$

with η_m as the mechanical efficiency and η_{ge} as the efficiency of the coupled generator and the electrical auxiliary equipment. The efficiency η is also defined by the relationship of the electrical power to the power potential in the wind:

$$\eta = \frac{P_{el}}{\frac{1}{2} \cdot \rho_a \cdot A \cdot v^3}. \tag{3.12}$$

3.3.4 Principles of Design

The issues discussed above can be summed up and related to the design of a wind energy converter through the following principles:

1. A high aerofoil form ratio c_A/c_R leads to a high tip to speed ratio and therefore a large power coefficient c_p.
   $\longrightarrow$ Modern converters with a good aerodynamic profile rotate quickly.
2. Simple profiles with low aerofoil form ratios have a small tip speed ratio. Therefore, the area of the rotor radius that is occupied by blades must be increased in order to increase the power coefficient.
   $\longrightarrow$ Slow rotating converters have poor aerodynamic profiles and a high number of blades.
3. The profile form ratio and the tip speed ratio have a considerably greater influence on the power coefficient than the number of blades.
   $\longrightarrow$ For modern converters with a good aerodynamic profile, the number of blades is not so important for a large power coefficient c_p.

3.4 Wind Variations

3.4.1 Wind Shear with Height

Assuming that the wind is blowing at 10 m/s at a height of 100 m, Fig. 3.6 shows how wind speeds vary in agricultural land with some houses and sheltering hedgerows with some 500 m intervals.

The fact that the wind profile shown in Fig. 3.6 is twisted towards a lower speed, as one moves closer to ground level, is usually called wind shear. Wind shear may also be important when designing wind turbines. It may be noticed that for a wind

Fig. 3.6 Variation in wind velocity with altitude

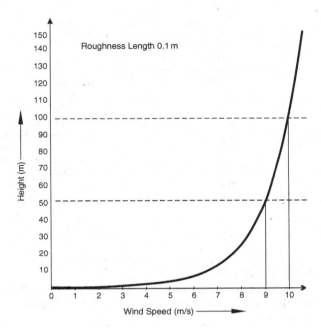

height of 50 m, the wind velocity is 9 m/s whereas for a 100 m height it is 10 m/s in our example case. With the help of the formula for power of wind discussed earlier in this chapter, it can be calculated that due to the dependence of power with the cube of wind speed, this increase of 1 m/s corresponds to about 30% difference in the power available from wind. It is also important to note that if a wind turbine with a hub height of 75 m and a rotor diameter of 50 m is considered, one can notice that the wind is blowing at 10 m/s when the tip of the blade is in its uppermost position (100 m height), and 9 m/s when the tip is in the bottom position (50 m height). This means that the forces acting on the rotor blade when it is in its top position are far larger than when it is in its bottom position.

The wind speed at a certain height above ground level is given by:

$$v = v_{ref} \ln(z/z_0)/\ln(z_{ref}/z_0)$$

where:

v	= wind speed at height z above ground level,
v_{ref}	= reference speed, i.e. a wind speed at height z_{ref},
$\ln(...)$	= natural logarithm function.,
z	= height above ground level for the desired velocity v,
z_{ref}	= reference height, i.e. the height at which wind speed is v_{ref},
z_0	= roughness length in the current wind direction.

Table 3.1 Roughness class and roughness lengths

Roughness class	Roughness length (m)	Energy index (%)	Landscape type
0	0.0002	100	Water surface
0.5	0.0024	73	Completely open terrain with smooth surface, e.g. airport runways, mowed grass
1	0.03	52	Open agriculture land without fencing and very scattered buildings, softly rounded hills
2	0.1	39	Agriculture land with some houses and 8 m tall sheltering hedgerows at a distance of approx. 500 m
3	0.4	24	Villages, small towns,
4	1.6	13	Very large cities with tall buildings and skyscrapers

A List of all major types of roughness classes and their typical roughness lengths is given in Table 3.1.

The above example is assuming that the wind is blowing at 9 m/s at 50 m height and the wind speed at 100 m height is to be calculated. If the roughness length is 0.1 m, then:

$$v_{ref} = 9 \text{ m/s}$$
$$z = 100$$
$$z_0 = 0.1$$
$$z_{ref} = 50 \text{ hence,}$$
$$v = 9 \ln(100/0.1)/\ln(50/0.1) = 10 \text{ m/s}$$

Average wind speeds are often available from meteorological observations measured at a height of 10 m. Hub heights of modern 600–1,500 kW wind turbines are usually 40–80 m and more. Using the above mentioned approach, one may calculate average wind speeds at different heights and roughness classes. It is to be noted that the results are not strictly valid if there are obstacles close to the wind turbine (or the point of meteorological measurement) at or above the specified hub height. It should be rather noted that there may be inverse wind shear on hilltops because of the hill effect, i.e. the wind speed may actually decline with increasing height during a certain height interval above the hilltop. A careful study of wind velocity and profile is therefore recommended before arriving at any conclusions about site.

3.4.2 Influence of Weibull Distribution

It is very important to understand the variation of wind speeds within the studied/measured time periods since it governs the energy output significantly. If one measures the wind speeds throughout a year, it can be noticed that in most areas

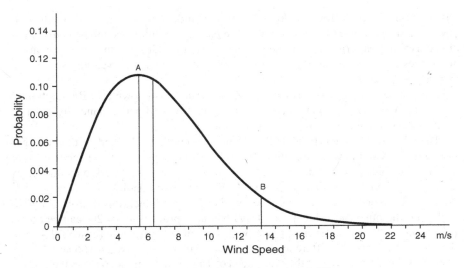

Fig. 3.7 Weibull distribution plot between wind velocity and probability

strong gale-force winds are rare, while moderate and fresh winds are quite common. The impact of such variation on energy output is described in this section.

The wind variation for a typical site is usually described using the so-called Weibull distribution, as shown in Fig. 3.7. This particular site has a mean wind speed of 7 m/s, and the shape of the curve is determined by a so called shape parameter of 2.

It can be realized that the shown graph is a probability distribution. The area under the curve is always exactly 1, since the probability that the wind will be blowing at some wind speed must be 100%.

Half of the area is to the left of the vertical black line at 6.6 m/s, and half is on its right hand side. The 6.6 m/s is called the median of the distribution shown above. From site-to-site, this median value would be differing. For this site, it means that half the time it will be blowing less than 6.6 m/s, the other half it will be blowing faster than 6.6 m/s. As it can be seen, the distribution of wind speed is skewed, i.e. it is not symmetrical. Sometimes there will be very high wind speeds, but they are very rare. Wind speeds of 5.5 m/s, on the other hand, are the most common ones. The statistical distribution of wind speed varies from place to place around the globe, depending upon local climate conditions, the landscape, and its surface.

The Weibull distribution may vary from site to site, both in its shape, and in its median value. There are two parameters that govern the shape of the Weibull distribution curve, namely, the scale parameter and the shape parameter. A higher value of the scale parameter means the distribution is spread over a wider range and the probabilistic average wind velocity has a higher value. A higher value of shape parameter (between 2 and 3) means the distribution is more skewed towards higher wind velocities, if the shape parameter is between 1 and 2, it means that the distribution is skewed towards lower velocities, indicating a higher probability of lower wind velocities. Of course, both these parameters influence the peak distribution curve, but one has major influence on the average value and the other primarily influences the skewness of the curve. For an exact feeling of curve, both should be considered together.

If the shape parameter is exactly 2, the distribution is known as a Rayleigh distribution. Wind turbine manufacturers often give standard performance figures for their machines using the Rayleigh distribution.

The reason why care is to be taken in connection with wind speeds is their energy content. The wind power varies with the cube of the wind speed. Design/selection of a wind turbine as per mathematical average wind speed may lead to an improper utilization of wind potential at the site. In Fig. 3.7, two conditions marked A and B, one corresponding to the peak velocity i.e. 5.5 m/s which has a probability of 11%, and the other the velocity of 14 m/s having a probability of 2% can be compared. Due to the dependence of power with cube of velocity, at these two velocities 5.5 m/s and 14 m/s, the difference between their cubes, i.e. between 166 and 2,744, is huge. Which means that at the high speed of 14 m/s, more than 16 times the power is available but only for 2% duration, whereas for a large duration of 11% only a small amount of power is available. If the wind turbine is selected as per 14 m/s wind velocity, it would remain much underutilized since the probability of the presence of smaller wind velocities is much higher. However, these figures are to be seen with respect to the energy delivered by the wind turbine and the time/day/month of delivery. Some countries such as Germany, where a feed-in tariff system exists, only the amount of power delivered is important, whereas, at some places the time of delivery may also be important. This analysis that affects the selection of a machine very much, can be done through the power curve of the machine being selected, which is discussed in Chap. 6.

Chapter 4
Components of a Wind Energy Converter

A wind energy converter has the following major components:

1. Rotor blades
2. Gearbox
3. Generator
4. Tower
5. Miscellaneous parts

All of these parts, their major design features and variations available on the market are discussed in this chapter. Besides these major components, there are other necessary parts, such as grid connection. Options for grid connection are covered in Chap. 6.

4.1 Rotor Blades

The basic principles behind the working of wings of airplanes and blades of wind turbines are quite common. However, since the wind turbines actually work in a very different environment with changing wind speeds and changing wind directions, there are special considerations that are not important in the design of airplane wings.

Figure 4.1 shows a photo of a rotor blade for a 1.5 MW wind converter. Figure 4.2 defines the four sides of a blade.

Due to the aerodynamic profile of blades, as discussed in Chap. 3, low pressure is created on the upper surface of a blade. This creates the lift, i.e. the force pulling upwards, the same principle that enables the plane to stay in the air. In case of the rotor blade of a wind mill, the lift is perpendicular to the direction of the wind.

Choosing profiles for rotor blades involves a number of compromises including reliable pitch control and stall characteristics, and the profile's ability to perform well even if there is some dirt on the surface. In countries where rains occur during the whole year, the dirt gets cleaned by the rain water, but in relatively dry countries like India, dirt may become a problem. Even in Germany, when there is only light rain, it may not be sufficient to whip-off the dirt from the blades. Under these

H.-J. Wagner, J. Mathur, *Introduction to Wind Energy Systems*, Green Energy and Technology, DOI 10.1007/978-3-642-02023-0_4, © Springer-Verlag Berlin Heidelberg 2009

Fig. 4.1 Rotor blade of a 1.5 MW wind turbine (photo: Wagner)

Fig. 4.2 Four sides of a rotor blade of a wind turbine

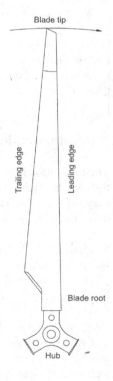

circumstances, the blades will have to be cleaned and possibly even polished again after 4–6 years, to ensure a smooth aerodynamic effect.

Most modern rotor blades on large wind turbines are made of glass fibre reinforced plastics, (GRP), i.e. glass fibre reinforced polyester or epoxy. Using carbon fibre or aramid (Kevlar) as reinforcing material is another possibility, usually such blades become necessary in large turbines in the range of 3–6 MW capacity. Wood, wood-epoxy, or wood-fibre-epoxy composites have not penetrated the market for rotor blades. Steel and aluminum alloys have problems of weight and metal fatigue respectively.

Large rotor blades typically consist of a metal walled circular root section with metal bushings with bolts or studs for mounting the blade to the hub. This root section is integrated into a continuously tapered longitudinal beam, the spar, which provides the stiffness and strength required to carry the wind load and the weight of the blade. Around the spar, the two aerodynamic-shaped shells, the suction side and the pressure side, form an optimized aerodynamic wing. The outer shells of the blade meet at the leading edge and the trailing edge. Inserted webs take up the torsional twist of the blade and help stabilize the blade against bending, shear loads and global buckling.

4.2 Gearboxes

The power from the rotation of the wind turbine rotor is transferred to the generator through the power train, i.e. through the main shaft, the gearbox and the high speed shaft. It is not appropriate to just drive the generator directly with the power from the main shaft unless the turbine is equipped with multi-pole generators that are discussed in the next chapter.

If an ordinary generator with two poles (which makes one pair of poles) was used without a gearbox, and the machine was directly connected to a 50 Hz AC (alternating current) three-phase grid, it would become necessary to turn the rotor at 50 revolutions per second i.e. 3,000 revolutions per minute (rpm). With a 50 m rotor diameter that would imply a tip speed of the rotor of far more than twice the speed of sound, a condition which cannot be accepted. Usually, a tip-speed of more than 100 m/s is not accepted across the industry. Another possibility of avoiding a gearbox is to build a slow-moving AC generator with a lot of poles. If a generator was to be connected directly to the grid, this would require a 300 pole generator to arrive at a reasonable rotational speed of 20 rpm.

With a gearbox, slowly rotating, high torque power obtained from the wind turbine rotor can be converted into high speed power, which would be required for the generator. The gearbox in a wind turbine usually does not "change gears". It normally has a single gear ratio between the rotation of the rotor and the generator. In some machines, the gear ratio is in the range of 30–200.

A typical issue for a gearbox is cooling and heating of the lubricant. In cold climatic conditions like in Germany, in winters, the lubricating oil freezes due to extreme low temperatures. Special heating arrangements are made for keeping the oil sufficiently warm. The situation in hot climates, like in India, is reverse. Arrangements are to be made to prevent the lubricating oil from becoming hot.

A recent development in the field of gearboxes for modern machines is the development of hydrodynamic gearboxes that are actually a combination of a conventional mechanical gearbox and a torque changer. There is a variable speed drive

inside and a constant drive outside. The main advantage of this type is continuously variable input but constant output speed adjustment over a wide range.
Prototypes have been successfully tested and this technology is very close to the
commercialization stage.

4.3 Generators

The wind turbine generator converts mechanical energy to electrical energy. Wind
turbine generators are a bit different, compared to other generating units ordinarily
attached to the electrical grid. One reason is that the generator has to work with a
power source (the wind turbine rotor) which supplies very fluctuating mechanical
power (torque).

4.3.1 Generating Voltage

In large wind turbines the voltage generated by the turbine is usually around 690 V
three-phase alternating current (AC). The current is subsequently sent through a
transformer next to the wind turbine (or inside the tower) to raise the voltage to
somewhere between 10,000 and 30,000 V, depending on the standard in the local
electrical grid. Manufacturers will supply both 50 Hz wind turbine models (for the
electrical grids in most of the world) and 60 Hz models (for the electrical grid e.g.
in America).

4.3.2 Generator Cooling System

Generators need cooling while they work. On most turbines this is accomplished by
encapsulating the generator in a duct, using a large fan for cooling by air, but a few
manufacturers use hydraulically cooled generators. Hydraulically cooled generators
may be built more compactly, but they require a heat exchanger (radiator) in the
nacelle to get rid of the heat from the liquid cooling system.

4.3.3 Generator Rotational Speed

The speed of a generator which is directly connected to a grid is constant, and
dictated by the frequency of the grid. The relationship between the rotational speed
of generator and frequency is governed by the formula:

$$\text{Rotation per second} = \frac{\text{Frequency per second } (50\,\text{Hz or } 60\,\text{Hz})}{2 \times \text{ number of poles}}. \qquad (4.1)$$

Table 4.1 Synchronous Generator Speeds in records per minute (rpm)

Pole number	No. of pole pairs	50 Hz	60 Hz
2	1	3,000	3,600
4	2	1,500	1,800
6	3	1,000	1,200
8	4	750	900
10	5	600	720
12	6	500	600
300	150	20	–
360	180	–	20

The number of poles means the number of coil sets in the stator of the generator which the electric power is generated in. Two poles make one pole pair. The number of poles will be decided by the construction of generator. Because of the fact that grid is operating with three-phases, the generator must also produce three phase alternating current.

When the number of pole pairs is doubled in the stator of a synchronous generator, the rotation will be reduced to half. The variations between rotation per minute and pole pairs are given in Table 4.1.

Often wind turbines that are with gearboxes, use generators with two or three pole pairs. The reasons for using these relatively high-speed generators are savings on size and cost. The maximum force (torque) a generator can handle depends on the rotor volume. The other type of wind turbines that do not use a gearbox but use multi-pole generators (explained in the next chapter), have a large number of poles.

The number of poles is kept less than 300 (say 84) due to weight considerations. As a result, the frequency of electricity generated is less than 50 Hz, which is transformed into 50 Hz before supply to the grid electronically.

The term "synchronous generator speed " (f/p, the ratio of frequency and number of pole pairs) thus refers to the speed of the generator when it is running synchronously with the grid frequency. It applies to synchronous generators . However, in the case of asynchronous (induction) generators it is equivalent to the idle speed of the generator. This means that the generator rotation must be a little bit higher than the synchronous speed e.g. in case of 2 pole pairs, 1,515 rpm for a synchronous speed of 1,500 rpm. The advantages and disadvantages of synchronous and asynchronous generators are further discussed in Chap. 5.

Wind turbines which use synchronous generators use electromagnets normally in the rotor which are excited by direct current from the electrical grid. Since the grid supplies alternating current, they first have to convert alternating current to direct current before sending it into the coil windings around the electromagnets in the rotor. The rotor electromagnets are connected to the current by using brushes and slip rings on the axle (shaft) of the generator.

Permanent magnet synchronous generators are tested in prototypes but not used so often despite their advantage of having a low weight. There are several reasons for this. One reason is the fear that permanent magnets could become demagnetized because of the mechanical forces. Another reason is that powerful magnets (made of rare earth metals, e.g. Neodymium) are quite expensive, although prices have dropped lately.

The key component of the basic version of an asynchronous generator is the squirrel cage rotor or simple cage rotor.

For understanding the operation of an asynchronous generator in basic version, consider manual cranking of this rotor around at exactly the synchronous speed of the generator, e.g. 1,500 rpm, for the 2-pole pair synchronous generator. Since the magnetic field rotates at exactly the same speed as the rotor, there would be no induction phenomena in the rotor, and it would not interact with the stator. The stator is connected to the grid and its field is rotating recording the grid frequency and pole numbers with 1,500 rpm. If the speed of rotor is increased above 1,500 rpm, in that case the rotor moves faster than the rotating magnetic field from the stator, in that condition, once again the stator induces a strong current in the rotor. The harder the rotor is cranked, the more power will be transferred as an electromagnetic force to the stator. The current in the rotor will create an electromagnetic field which generates electricity in the stator, which is fed into the electrical grid.

It is the rotor that makes the asynchronous generator different from the synchronous generator. The rotor consists of a number of copper or aluminum bars which are connected electrically by copper or aluminum end rings. The rotor is placed in the middle of the stator which is directly connected to the three-phases of the electrical grid. The clever thing about the cage rotor is that it adapts itself to the number of poles in the stator automatically. The same rotor can, therefore, be used with a wide variety of pole numbers.

4.3.4 Two Speed, Pole Changing Generators

Some manufacturers fit their turbines with two generators, a small one for periods of low winds, and a large one for periods of high winds. A more common design on newer machines is pole changing generators, i.e. generators which (depending on how their stator windings are connected) may run with a different number of pole pairs, and thus a different rotational speed.

Whether it is worthwhile to use a switching of pole pairs in the generator depends on the local wind speed distribution, and the extra cost of the pole pairs changing generator compared to the price the turbine operator gets for the electricity. However, due to the problems related to a drop in grid voltage, the practice of switching the poles has been discontinued in Europe.

4.4 Towers

The tower of the wind turbine carries the nacelle and the rotor blades. The tower needs to be as tall as possible, because the wind speed increases with height. However, the height is optimized by analyzing the cost of the increase in tower height and the gain in energy output due to increased wind velocity at a greater height. Towers for large wind turbines may be either tubular steel towers, lattice towers, or concrete towers. Guyed pole towers are only used for small wind turbines.

(a) *Tubular steel towers.* Most large wind turbines are delivered with tubular steel towers, which are manufactured in sections of 20–30 m with flanges at either end, and bolted together on the site. The towers are conical (i.e. with their diameter increasing towards the base) in order to increase their strength and to save materials at the same time.

(b) *Lattice towers.* Lattice towers are manufactured using welded steel profiles. The basic advantage of lattice towers is cost, since a lattice tower requires only half as much material as a freely standing tubular steel tower with a similar stiffness.

The advantage of the low cost of lattice towers exists only in those countries where the labor cost is cheap e.g. in India. In countries like Germany, where the labor costs are high, the advantage of reduced material cost gets balanced by increased labor costs for assembling the tower. Another disadvantage of lattice towers is their visual appearance, although that issue is clearly debatable. Be that as it may, for aesthetic reasons lattice towers have almost disappeared from being used for large, modern wind turbines. These towers are used only in small machines i.e. below the MW size.

(c) *Concrete towers.* Recently, with a further increase in the height of towers for more than 100 m hub height, due to the increased cost of steel in towers, manufacturers have started to bring concrete towers in place of steel towers on the market. In future, therefore, there may be an additional type of tower on the wind energy market.

(d) *Guyed pole towers.* Many small wind turbines are built with narrow pole towers supported by guy wires. The advantage lies in saving weight and, consequently, cost. The disadvantage lies in a difficult access around the towers which makes them less suitable in farm areas. Such towers are used only in kW size machines.

(e) *Hybrid tower solutions.* Some towers are built by using different combinations of the techniques mentioned above. It may be a hybrid between a concrete tower (down) and steel (upward) or between a lattice tower and a guyed tower.

Generally, it is an advantage to have a tall tower in areas with a high terrain roughness, since the wind speeds increase with height above the ground, as discussed in previous chapters. Lattice towers and guyed pole towers, due to a reduced surface area as compared to tubular steel towers, have the advantage of giving less wind shade than a massive tower. However, the turbulence created by each element of a lattice structure adversely affects the energy output.

4.5 Miscellaneous Components

Large wind turbines are equipped with a number of safety devices to ensure safe operation during their lifetime. A brief description of such devices is given below:

(a) *Yaw mechanism.* Almost all horizontal axis wind turbines use forced yawing, i.e. they use a mechanism which uses electric motors and gearboxes to keep the rotor plane perpendicular to the direction of the wind. Almost all manufacturers of upwind machines prefer to apply brakes to the yaw mechanism whenever it is unused. The yaw mechanism is activated by the electronic controller which checks the position of the wind vane on the turbine several times per minute, whenever the turbine is running.

Cables carry the current from the wind turbine generator down through the tower. Occasionally, a wind turbine may look as if it had gone berserk, yawing continuously in one direction for about any revolutions. The cables, however, will become more and more twisted if the turbine by accident keeps yawing in the same direction for a long time. The wind turbine is, therefore, equipped with a cable twist counter, which tells the controller that it is time to untwist the cables. The turbine is also equipped with a pull switch which gets activated if the cables become too twisted.

Besides the role of tracking the wind direction, the yaw mechanism plays another very important role by connecting the tower with the nacelle. Due to the operation of gears, it needs regular lubrication for smooth operation.

(b) *Brakes*. Braking action may be required for several reasons. There are several types of braking a turbine rotor: aerodynamic brakes , electro brakes and mechanical brakes. In case of aerodynamic braking, the blade is turned in such a direction that the lift effect which causes rotation, does not appear. This concept is explained in the section of "stall control" in Chap. 6. In case of electro-magnetic braking, energy produced by the generator of a wind turbine is dumped into a resistor bank, thereby converting it into heat.

Another type of braking is conventional mechanical braking for which disc brakes are provided in the nacelle. A mechanical drum brake or disk brake is also used to hold the turbine at rest for maintenance. Such brakes are also applied after blade furling and electro braking have reduced the turbine speed, as the mechanical brakes would wear quickly if used to stop the turbine from full speed.

In large wind turbines, normally there is a combination of at least two brakes, most turbines use aerodynamic brakes together with mechanical braking or even also with an electro braking system.

(c) *Over-speed protection*. In addition to braking explained above, it is essential that wind turbines stop automatically in case of malfunction of a critical component, for example, if the generator overheats or is disconnected from the electrical grid and the rotor will start accelerating rapidly within a matter of seconds. In such a case it is essential to have an over-speed protection system. Most companies provide wind turbines with two independent fail-safe brake mechanisms to stop the turbine.

(d) *Vibration measurement*. Excessive vibrations may be one of the most dangerous causes for failure of a wind turbine as well as any other component. Vibrations cause fatigue loading of components causing early failure. An additional side effect of vibration is the generation of noise. One of the classical, and most simple safety devices in a wind turbine is the vibration sensor. It simply consists of a ball resting on a ring. The ball is connected to a switch through a chain. If the turbine starts shaking, the ball will fall off the ring and switch the turbine off. Modern turbines have permanently installed vibration monitoring systems which in the case of excessive vibrations decide on the operation of the machine with the help of microprocessors.

(e) *Anemometer*. Before actuating the controlling actions, measurement of wind speed and direction is necessary in every wind turbine. It is usually done using a cup anemometer . The cup anemometer has a vertical axis and three cups which capture the wind. The number of revolutions per minute is registered electronically. Normally, the anemometer is fitted with a wind vane to detect the wind direction. Other anemometer types include ultrasonic or laser anemometers which detect the phase shifting of sound or coherent light reflected from the air molecules. Hot wire anemometers detect the wind speed through minute temperature differences between wires placed in the wind and in the wind shade (the leeward side). The advantage of non-mechanical anemometers may be that they are less sensitive to icing in cold countries and jamming due to dust in hot locations.

(f) *Lubrication system*. Since the wind turbine has moving parts, lubrication is always a major concern. There are two types of lubrication systems . With the first type, there is a central lubrication pump which sends lubricant to all bearings through tubes laid for this purpose. The other system has a pressurized container of lubricant for each bearing which directly lubricates the bearings.

(g) *Foundation*. Foundation plays a very important role in stabilizing the wind turbine. Due to the large height, heavy weight at the nacelle and large rotor area which faces wind forces, the role of a foundation becomes very important. While in case of on-shore installations, the type of foundation depends upon the nature of the soil, in case of off-shore turbines, it becomes an even more serious issue. Its designs tends to rely on technology used by the oil and gas industry in some cases. There are different major types of foundations used in case of off-shore wind turbines. The first is the mono pile foundation (see Fig. 4.3) that consists of a steel pile which is

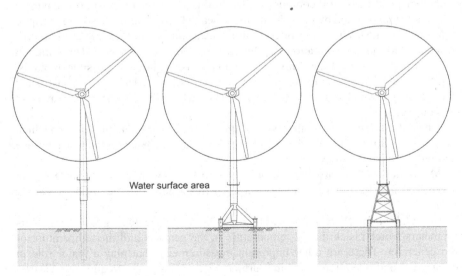

Fig. 4.3 Three foundations for off-shore wind turbines. From left to right: Monopile, Tripod, Jacked

Fig. 4.4 Tripod foundation for the windpark Alpha Ventus in Germany (photo: Große Böckmann)

driven approximately 10–12 m into the seabed. The second one is the Tripod foundation (Fig. 4.4). The piles on each end are typically driven about 10 m into the seabed, depending on soil conditions. The third is a so called jacked . It is having a lattice tower also, fixed by piles driven in the seabed. The fourth is a so called tripile solution (Fig. 4.5). With the tripile solution the advantages of the simpler monopole construction and transportation can be used, but the necessary load distribution on tops must designed carefully, concerning consequences of possible motions. The next one is the gravity foundation . It consists of a large base constructed from concrete which rests on the seabed. The turbine is dependent on gravity to remain erect.

Under development is another type of foundation, a swimming platform on which the wind tower is installed. The platform will swim a little bit under the water surface and will be fixed by guy wires on sea ground.

Monopiles and gravity foundation seems to be used in water depth less than about 20 m. The other foundation constructions will also be used in deeper water.

(h) *Other parts*. There are many other sensors in the nacelle, e.g. electronic thermometers which check the oil temperature in the gearbox and the temperature of the generator. There are other important measurements that play a major role in connection and operation of a wind turbine with grid: voltage-frequency measurement, phase-angle measurement, revolution measurement, electric power supply measurement. There are self controlling systems with a telecommunication link to a

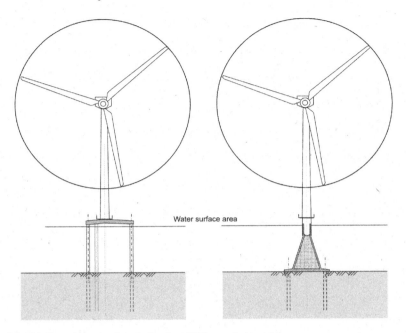

Fig. 4.5 Tripile and gravity foundation for off-shore wind turbines

common service point. All of these make the wind turbine more efficient and available for a maximum time for the economic generation of power. The components related to grid connection are discussed in Chap. 6.

Chapter 5
Design Considerations

When finalizing the specifications of wind turbines for installation at any site, a lot of technological options are found to be available. These options along with a discussion on their suitability in various situations are presented in this chapter.

5.1 Rotor Area of Turbines

The question of small or large turbines pertains to the decision related to the area of rotor. A larger rotor with compatible generator would be called a large turbine.

5.1.1 Reasons for Choosing Large Turbines

1. There are economies of scale in wind turbines, i.e. larger machines are usually able to deliver electricity at a lower cost than smaller machines. The reason is that the cost of foundations, road building, electrical grid connection, plus a number of components in the turbine (the electronic control system etc.) are somewhat independent of the size of the machine.
2. Larger machines are particularly well suited for offshore wind power. The cost of foundations does not rise in proportion to the size of the machine.

5.1.2 Reasons for Choosing Smaller Turbines

1. The local electrical grid may be too weak to handle the electricity output from a large machine. This may be the case in remote parts of the electrical grid with a low population density and little electricity consumption in the area.
2. The cost of using large cranes, and building a road strong enough to carry the turbine components may make smaller machines more economic in some areas.
3. Aesthetic landscape considerations are dependent upon the perception of local people. Some people like to see larger machines and attach them with their own identity. Another set of people may not want to see any extraordinary large structures in the natural landscape. Speed of rotation also plays an important role

H.-J. Wagner, J. Mathur, *Introduction to Wind Energy Systems*, Green Energy and Technology, DOI 10.1007/978-3-642-02023-0_5, © Springer-Verlag Berlin Heidelberg 2009

in perception about landscape, since large machines have a much lower rotational speed, which means that one large machine really does not attract as much attention as many small, fast moving rotors.

In countries like Germany, where the availability of a strong grid to accept power is not a problem, the question of a small or large rotor area becomes insignificant. There is a clear cut tendency to prefer larger machines due to the economic advantage with a large rotor diameter. However, the selection between large and small machines becomes an important criterion in developing countries like India where the selection of larger machines may become inappropriate in the absence of availability of a strong grid.

5.2 Number of Blades

In modern wind turbines, use of an even number of rotor blades is avoided. The most important reason is the stability of the turbine. A rotor with an odd number of rotor blades (and at least three blades) can be considered to be similar to a disc when calculating the dynamic properties of the machine. A rotor with an even number of blades will give stability problems for a machine with a stiff structure. The reason is that at the very moment when wind is coming from the front, the uppermost blade bends backwards, because it gets the maximum power from the wind. If an even number of blades is used, at that time the lowermost blade passes into the wind shade in front of the tower. This combination makes the dynamic balancing of rotor more well-balanced.

Most modern wind turbines, therefore, are three-bladed designs. The vast majority of the turbines sold in world markets have this design. Selection of the number of blades depends upon the wind profile.

Besides the above mentioned disadvantage, two-bladed wind turbine designs have the advantage of saving the cost of one rotor blade and its weight, of course. However, they tend to have difficulties in penetrating the market, partly because they require higher rotational speed to yield the same energy output. This is a disadvantage both in regard to noise and visual intrusion. Lately, several traditional manufacturers of two-bladed machines have switched to three-bladed designs.

One-bladed wind turbines also exist as prototypes, and indeed, they save the cost of another rotor blade. One-bladed wind turbines are not very widespread commercially, however, because the same problems that are mentioned with the two-bladed design apply to an even larger extent to one-bladed machines. In addition to the higher rotational speed, and the noise and visual intrusion problems, they require a counterweight to be placed on the other side of the hub from the rotor blade in order to balance the rotor. This obviously negates the savings on weight compared to a two-bladed design.

Two- and one-bladed machines require a more complex design with a hinged (teetering hub) rotor. The rotor has to be able to tilt in order to avoid too heavy

Fig. 5.1 Power coefficients and tip-speed ratio of different types of rotors (A: Savonius rotor, B: Multi blade rotor, C: Four blade rotor, D: Three blade rotor, E: Two blade rotor)

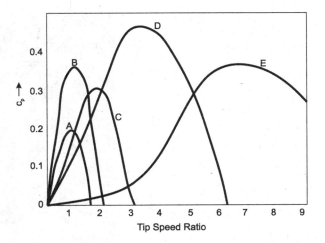

shocks to the turbine when a rotor blade passes the tower. The rotor is, therefore, fitted onto a shaft which is perpendicular to the main shaft, and which rotates along with the main shaft. This arrangement may require additional shock absorbers to prevent the rotor blade from hitting the tower.

Due to mechanical problems, two blade and single blade designs have not become very popular. Additionally, it is understood that the three blade design is psychologically more acceptable to human perception as compared to the two or single blade design. Figure 5.1 shows the range of power coefficient and tip speed ratio for different types of rotors. It shows that as compared to the three blade rotors, the two blade rotors need a higher tip-speed ratio for operating at the same value of power coefficient. This means that for a given wind velocity, a two blade rotor has to rotate faster as compared to a three blade rotor.

5.3 Horizontal or Vertical Axis Turbine

Considering the present status of trade, the choice between horizontal axis and vertical axis wind turbines is almost theoretical and goes in favor of horizontal axis turbines. However research is going on to bring back vertical axis turbines into practice, which may be possible in future.

(a) *Horizontal axis wind turbines.* Most of the technology described on these pages is related to horizontal axis wind turbines. The reason is simple: All grid-connected commercial wind turbines today are built with a propeller-type rotor on a horizontal axis (i.e. a horizontal main shaft). The purpose of the rotor, of course, is to convert the linear motion of the wind into rotational energy that can be used to drive a generator. The same basic principle is used in a modern water turbine, where the flow of water is parallel to the rotational axis of the turbine blades.

Fig. 5.2 A 4.2 MW vertical
axis Darrieus wind turbine
(*source:* www.reuk.co.uk)

(b) *Vertical axis wind turbines*. Figure 5.2 shows the biggest prototype of a vertical axis wind turbine, with a 100 m rotor diameter and a capacity of 4.2 MW at Cap Chat, Québec, Canada. The machine is no longer operational. It was operational in the period 1983–1992. The only vertical axis turbine which has ever been manufactured commercially at any volume is the Darrieus machine, named after the French engineer Georges Darrieus who patented the design in 1931. It was manufactured by the U.S. company FloWind which stopped its production in 1997. The Darrieus machine is characterized by its C-shaped rotor blades. It is normally built with two or three blades. Another type of vertical axis machine is the machine with H-rotor. The name is derived from the shape "H" of its rotor (see Fig. 5.3). One more type of vertical axis machine consists of Savonius rotors which are primarily used for measuring weather conditions.

The basic theoretical advantages of a vertical axis machine are:

1. The generator, gearbox etc. can be placed on the ground, and may not need a tower for the machine.
2. A yaw mechanism is not needed to turn the rotor against the wind.

The basic disadvantages are:

1. Wind speeds are very low close to ground level, so the need of a tower is eliminated, but the wind speeds will be very low on the lower part of the rotor.

Fig. 5.3 H-rotor in Germany
(photo: Große Böckmann)

2. The overall efficiency of the vertical axis machines is less than in the case of a horizontal axis.
3. The machine is not self-starting, e.g. a Darrieus machine will need a "push" before it starts. This is only a minor inconvenience for a grid connected turbine, since the generator may be used as a motor drawing current from the grid to start the machine.
4. The machine may need guy wires to hold it up, but guy wires are impractical in heavily farmed areas.

5.4 Upwind or Downwind Turbine

Upwind machines are those machines that have the rotor facing the wind. In these machines the wind meets the rotor first and then leaves from the direction in which the nacelle is located. Downwind machines have the rotor placed on the leeward side of the tower; this means the nacelle comes first in the path of the wind and then the blades, as shown in Fig. 5.4.

(a) *Upwind machines*. The basic advantage of upwind designs is that one avoids the wind shade behind the tower. By far the vast majority of wind turbines have this design. On the other hand, there is also some wind shade in front of the tower,

Fig. 5.4 Upwind and
downwind type wind turbines

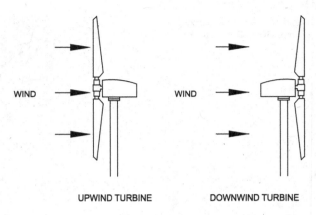

i.e. the wind starts bending away from the tower before it reaches the tower itself, even if the tower is round and smooth. Therefore, each time the rotor passes the tower, the power from the wind turbine drops slightly. The basic drawback of upwind designs is that the rotor needs to be placed at some distance from the tower. In addition, an upwind machine needs a yaw mechanism to keep the rotor facing the wind.

(b) *Downwind machines*. They have the theoretical advantage that they may be built without a yaw mechanism, if the rotor and nacelle have a suitable design that makes the nacelle follow the wind passively. For large wind turbines this is a somewhat doubtful advantage, since for optimal energy efficiency of wind energy converters, the yaw control must be applied very accurately.

Another advantage of the downwind design is that the rotor may be made more flexible. This is an advantage in regard to weight and the structural dynamics of the machine, i.e. the blades will bend at high wind speeds, thus taking part of the load off the tower.

The basic drawback is the fluctuation in the wind power due to the rotor passing through the wind shade of the tower. This may give more fatigue loads on the turbine than with an upwind design.

Out of these two options, upwind machines are more commonly used due to an increased energy output and hence their being more economical.

5.5 Load Considerations for Turbine Selection

Wind turbines are subject to fluctuating winds, and hence fluctuating forces. This is particularly the case if they are located in a very turbulent wind climate. Components which are subject to repeated bending, such as rotor blades, may eventually develop cracks which ultimately may make the component break. Metal fatigue is a well known problem in many technical goods. Due to combined reasons of fatigue and mass, metal is, therefore, generally not preferred as a material for rotor blades. When designing a wind turbine it is extremely important to calculate in advance how the different components will vibrate, both individually, and jointly. It is also important to calculate the forces involved in each bending or stretching of a component. This is the subject of structural dynamics, where mathematical computer models have been developed that analyse the behavior of an entire wind turbine. These models are used to design the machines safely.

As an example, a tall wind turbine tower will have a tendency to swing back and forth, say, every three seconds. The frequency with which the tower oscillates back and forth is also known as the eigenfrequency of the tower. The eigenfrequency depends on the height of the tower, the thickness of its walls, the type of steel, and the weight of the nacelle and rotor. Now, each time a rotor blade passes the wind shade of the tower, the rotor will push slightly less against the tower.

If the rotor turns with a rotational speed such that a rotor blade passes the tower each time the tower is in one of its extreme positions, then the rotor blade may either dampen or amplify (reinforce) the oscillations of the tower. The rotor blades themselves are also flexible, and may have a tendency to vibrate, say, once per second. Consequently, it is very important to know the eigenfrequencies of each component in order to design a safe turbine that does not oscillate out of control.

5.6 Wind Turbines: With or Without Gearbox

(a) *Design with gearbox.* The principle of a design of a wind turbine with gearbox, is shown in Fig. 5.5 and 5.6. The main aspect of this design is the split shaft system, where the main shaft turns slowly with the rotor blades and the torque is transmitted through a gearbox to the high-speed secondary shaft that drives the few-pole pair generator. The transmission of torque to the generator is shut off by means of a large disk brake on the main shaft. A mechanical system controls the pitch of the blades, so pitch control can also be used to stop the operation of the converter, e.g. in stormy conditions. The pitch mechanism is driven by a hydraulic system, with oil as the popular medium. For constructions without a main brake, each blade has its pitch angle controlled by a small electric motor.

The gearbox concept was in many cases accompanied by an insuffcient life time because of failure of gearboxes. After many years of operational experiences and a lot of research and development activities it got solved.

(b) *Design without gearbox.* Some companies, e.g. the German company Enercon, designed another converter type without gearbox. The scheme of such a converter is shown in Fig. 5.7, where the main design aspects can be clearly seen. This design has just one stationary shaft. The rotor blades and the generator are both mounted on this shaft. The multi-pole generator is in the form of a large spoked wheel with e.g. forty-two pole pairs around the outer circumference and stators mounted on a stationary arm around the wheel. The wheel is fixed to the blade apparatus, so it rotates slowly with the blades. Therefore, there is no need for

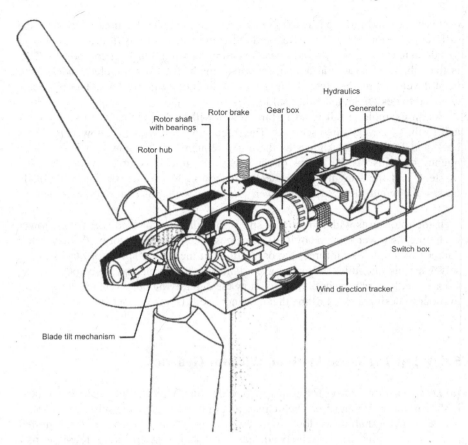

Fig. 5.5 Wind turbine with a gearbox

a gearbox, rotating shafts or a disk brake. This minimizing of rotating parts reduces maintenance and failure possibilities and simplifies the maintenance and production of the converter. The price for this advantages is a high nacelle mass caused by the high copper content of multi-pole generator.

(c) *Multibrid design*. Both designs explained above have one disadvantage each. The design with gearbox has the disadvantage of losses during transmission of power and high speed drive required for connecting the generator, while the gearless design has the disadvantage of increased weight of the nacelle due to an increased number of poles. The objective of the multibrid design is to overcome disadvantages of both previous designs. In this design, a new combination of approved techniques has been used. It is operated in the M5000 design of the research and development company Multibrid (a registered name) and was developed especially for offshore use. This design is a combination between a special gearbox and a multi-pole generator. There is no high-speed transmission included in the gearbox, which is critical to failure. The selection of transmission achieves a high transmission allowing the

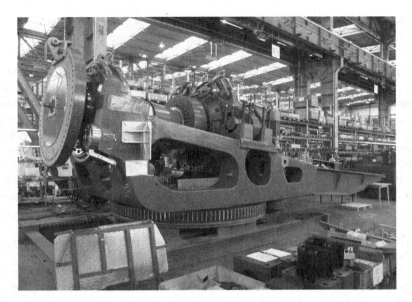

Fig. 5.6 Assembling of a wind turbine with gearbox by NORDEX company (photo: Nordex)

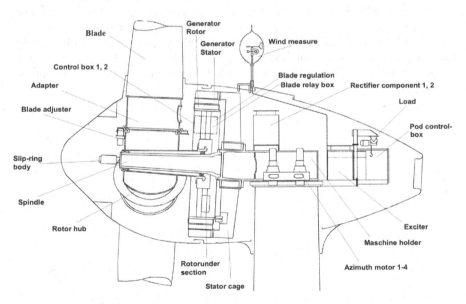

Fig. 5.7 Wind turbine without gearbox (Design of ENERCON company)

use of a generator with up to 150 rpm. Additionally, in the prototypes, a permanent magnet-excited synchronous generator with water-cooling, high efficiency and wide speed range is used. Due to the compact design, the construction has a reduced weight.

5.7 Requirement of Grid, Synchronous or Asynchronous Generators

The main electrical grid has a constant frequency, e.g. 50 Hz, and a constant phase angle. Therefore, a wind energy converter must produce electricity with the same constant values in order to integrate with electricity in the main grid. The input energy of a wind energy converter is proportional to the wind speed, but the wind speed is never constant. Each wind speed has a corresponding rotor rotation speed at which the maximum power is produced, as shown in Fig. 5.8. This maximum occurs at different rates of rotation for different wind speeds. However, the rate of rotation must be kept constant in order to achieve the required constant output frequency, e.g. by 18.5 rpm in the figure. Solutions to this problem of maximizing the power output of converters by variable speeds and constant frequency are discussed in connection with the generator.

With respect to their working principle, there are two choices for generators: synchronous generators or asynchronous generators.

(a) *Synchronous generators.* All generators use a rotating magnetic field for the generation of electricity. The more force (torque) is applied and the rotational speed is kept constant, the more electricity is generated, but the synchronous generator will still run at the same speed dictated by the frequency of the electrical grid.

If the generator is disconnected from the main grid, however, it will have to be cranked at a constant rotational speed in order to produce alternating current with a constant frequency. The synchronous generator is generating power with constant frequency. Consequently, with this type of generator it is normally recommended to use an indirect grid connection for being able to operate it with different rotational speeds.

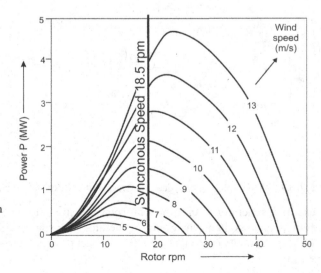

Fig. 5.8 Change in optimum rotation with wind speed for maximum power (synchronous speed 18.5 rpm)

Fig. 5.9 Grid connection of synchronous generators (SG) with full inverter system

a) Grid connection for synchronous generators with gearbox

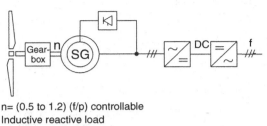

n= (0.5 to 1.2) (f/p) controllable
Inductive reactive load
Controllable reactive power

b) Grid connection for synchronous generators without gearbox

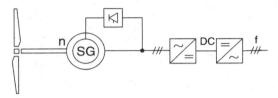

n= (0.5 to 1.2) (f/p) controllable
Inductive reactive load
Controllable reactive power

Figure 5.9 shows the connection of a wind turbine with synchronous generator to the grid. The synchronous generator operates with a variable speed of rotation, which means that it will produce electric power with variable frequencies. Due to a DC link, this will be transformed to AC power with the required frequency (usually the grid frequency). Use of DC has no link with the use of design having gearbox or without gearbox as shown in Fig. 5.9a, b for both cases.

(b) *Asynchronous (induction) generators*. The other option for choosing a generator is a three phase asynchronous generator, also called an induction generator to generate alternating current. This type of generator is widely used in the wind turbine industry and in small hydropower units. The unique feature of this type of generator is that it was really originally designed as an electric motor. One reason for choosing this type of generator is that it is very reliable, and tends to be comparatively inexpensive with the basic design of a cage rotor. The generator also has some mechanical properties which are useful for wind turbines such as a generator slip, and a certain overload capability.

The speed of the asynchronous generator must be little higher than the synchronous speed. In practice, the difference between the rotational speed at peak power and at idle is very small, up to 8%. This difference in per cent of the synchronous speed is called the generator's slip. Thus, a 4-pole generator will run idle at 1,500 rpm if it is attached to a grid with a 50 Hz current. If the generator is producing at its maximum power, lets say with 1% slip, it will be running at 1,515 rpm.

Fig. 5.10 Grid connection of
asynchronous generators
(ASG)

a) Direct grid connection

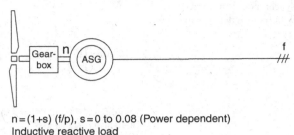

n = (1+s) (f/p), s = 0 to 0.08 (Power dependent)
Inductive reactive load

b) Dynamic slip control

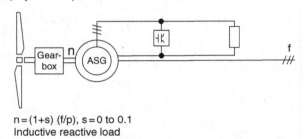

n = (1+s) (f/p), s = 0 to 0.1
Inductive reactive load

c) Double excited asynchronous generator

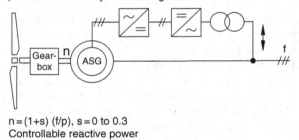

n = (1+s) (f/p), s = 0 to 0.3
Controllable reactive power

It is a very useful mechanical property that the generator will increase or decrease
its speed slightly if the torque varies. This means that there will be less tear and wear
on the gearbox due to lower peak torque. This is one of the most important reasons
for using an asynchronous generator rather than a synchronous generator on a wind
turbine which is directly connected to the electrical grid. The rotation of such direct
grid connected asynchronous generators varies in the range of 100% of synchronous
speed to 108% as shown in Fig. 5.10a.

In the section about the synchronous generator it has been discussed that it could
run as a generator without connection to the public grid. An asynchronous generator
is different, because it requires the stator to be magnetized from the grid before it
works. An asynchronous generator can also run in a stand alone system, however,

if it is provided with capacitors, which supply the necessary magnetization current. It also requires that there is some residual magnetism in the rotor iron, i.e. some leftover magnetism when the turbine is started. Otherwise a battery and power electronics or a small diesel generator is required to start the system. But it cannot run with constant frequency because the frequency is, as mentioned before, dependent on the rpm, which is dependent upon the generated electric power. In addition, an asynchronous generator can provide a limited amount of reactive power to the grid.

As discussed in the previous section, the motor (or generator) slip in an asynchronous (induction) machine is usually very small for reasons of efficiency, so the rotational speed will vary by up to 8% between idle and full load. The slip, however, is a function of the resistance in the rotor windings of the generator. So, one way of varying the slip is to vary the resistance in the rotor. In this way one may increase the generator slip to e.g. 10%. This can be done by having a wound rotor, i.e. a rotor with copper wire windings which are connected in a star, and connected with external variable resistors, plus an electronic control system to operate the resistors. The connection has usually been done with brushes and slip rings, which is a clear drawback over the elegantly simple technical design of a cage wound rotor machine. It also introduces parts which wear down in the generator. The rotation of such an asynchronous generator varies in the range of 100–110% of synchronous speed as shown in Fig. 5.10b.

Modern machines nowadays have double excited asynchronous generators as shown in Fig. 5.10c. In this case the generator is equipped with a winding in rotor and is excited with AC having variable frequency. This AC is produced by a DC link which is fed by the grid. This type of generators has an extended slip and is able to operate with rotational speed in the range of 100–120% of the synchronous speed. A further advantage of such generators is that they can supply more reactive power to the grid as compared to the other two types discussed above.

5.8 Issue of Noise and Its Control

The loudness or strength of a noise or sound signal is measured with the unit "decibel" abbreviated as "dB". One decibel is considered the smallest difference in sound level that the human ear can distinguish. Created in the early days of telephony as a way to measure cable and equipment performance and named after Alexander Graham Bell, decibels (dB) are a relative measurement derived from two signal levels: a reference input level and an observed output level. If a reference input level is used, the unit is often written as dB(A). A decibel is the logarithm of the ratio of the two levels. This means that if the dB level is doubling, the loudness level would be increasing by 10 times. A whisper is about 20 dB. A normal conversation is typically from 60 to 70 dB, and a noisy factory from 90 to 100 dB. Loud thunder is approximately 110 dB, and 120 dB borders on the threshold of pain.

The energy in sound waves (and thus the sound intensity) will drop with the square of the distance to the sound source, which is popularly known as the inverse square law. In other words, if one moves 200 m away from a wind turbine, the sound level will generally be one quarter of what it is 100 m away. To give an example: At one rotor diameter distance (43 m) from the base of a wind turbine emitting 100 dB(A) there will be generally a sound level of 55–60 dB(A) like a conversation. Four rotor diameters (170 m) away there will be 44 dB(A), corresponding to a quiet living room in a house. At 6 rotor diameters (260 m) it would be some 40 dB(A).

Sound emissions from wind turbines may have two different origins: Mechanical noise and aerodynamic noise. These two are explained below.

Mechanical noise, i.e. metal components moving or knocking against each other may originate in the gearbox, in the drive train (the shafts), and in the generator of a wind turbine.

When going by car, plane, or train, one may experience how resonance of different components, e.g. in the dashboard of a car or a window of a train may amplify noise. An important consideration, which enters into the turbine design process today, is the fact that the rotor blades may act as membranes that may retransmit noise vibrations from the nacelle and tower.

While designing, with the help of computer models it is ensured that the vibrations of different components do not interact to amplify noise. For example the chassis frame of the nacelle, on some of the wind turbines today, has some odd holes which are drilled into the chassis frame for no apparent reason. These holes were precisely made to ensure that the frame will not vibrate in step with the other components in the turbine.

When the wind hits different objects at a certain speed, it will generally start making a sound. If it hits the leaves of trees and bushes, or a water surface it will create a random mixture of high frequencies, often called white noise. The wind may also set surfaces in vibration, as sometimes happens with parts of a building, a car or even an engine-less glider plane. These surfaces in turn emit their own sound. If the wind hits a sharp edge, it may produce a pure tone.

In case of wind turbines, wind hits the rotor blades. Rotor blades make a slight swishing sound which may be heard close to a wind turbine at relatively low wind speeds. For their operation, rotor blades must split the wind on both sides of the blade so that energy can be transferred to the rotor. In the process they cause some emission of white noise. If the surfaces of the rotor blades are very smooth (which indeed they must be for aerodynamic reasons), the surfaces will emit a minor part of the noise. Most of the noise will originate from the trailing (back) edge of the blades. Careful design of trailing edges and very careful handling of rotor blades while they are mounted, have become routine practice in the industry for the purpose of controlling aerodynamic noise. Other things being equal, sound pressure will increase with the sixth power of the speed of the blade relative to the surrounding air. It can, therefore, be noticed that modern wind turbines with large rotor diameters have a very low rotational speed.

Since the tip of the blade moves substantially faster than the root of the blade, great care is taken about the design of the rotor tip. On close look at different rotor blades, it can be discovered that there has been substantial change in their geometry over time, as more and more research in the area has progressed. The research is also done for performance reasons, since most of the torque (rotational moment) of the rotor comes from the outer part of the blades. In addition, the airflows around the tip of rotor blades are extremely complex, compared to the airflow over the rest of the rotor blade.

One simple way of controlling noise is to reduce the tip-speed ratio by designing the rotor accordingly. An extension of this approach is sometimes used in Germany, by using different rotational speeds in day and night time due to different maximum permissible levels of noise. Many turbines can be seen with specially designed winglets at the tip of blades that help in reducing the noise, too, by controlling the turbulence at the tip.

Chapter 6
Operation and Control of Wind Energy Converters

After having discussed the appropriate wind energy system design, it is equally important to discuss its optimal operation and control of performance. The present chapter covers such aspects.

6.1 Power Curve and Capacity Factor

6.1.1 Power Curve

The power curve of a wind turbine is a graph that indicates which electrical power output will be available at different wind speeds. Figure 6.1 shows the shape of a theoretical power curve of a wind turbine.

Below the main features of any power curve are described:

(a) *The cut in wind speed (v_c).* Usually, wind turbines are designed to start running at wind speeds somewhere around 3–5 m/s. This is called the cut in wind speed. Below this speed of wind, the energy in wind is not sufficient to overcome the inertia of the rotor; hence, the machine does not produce any power below this speed of wind.

(b) *The cut out wind speed (v_f).* The wind turbine will be programmed to stop at high wind speeds above, say, 25 m/s, in order to avoid damaging the turbine or its surroundings. The stop wind speed is called the cut out wind speed.

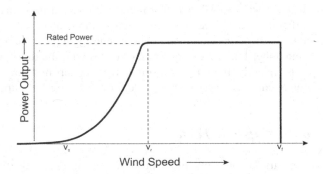

Fig. 6.1 Theoretical power curve of a wind turbine

H.-J. Wagner, J. Mathur, *Introduction to Wind Energy Systems*, Green Energy and Technology, DOI 10.1007/978-3-642-02023-0_6, © Springer-Verlag Berlin Heidelberg 2009

(c) *Rated wind speed* (v_r). The "rated wind speed" is the wind speed at which the "rated power" is achieved. This value for megawatt size turbines is about 12–15 m/s, and it corresponds to the point at which the conversion efficiency is near its maximum. The power output above the rated wind speed is mechanically or electrically maintained at a constant level, because the high output would destroy the equipment.

(d) *Survival speed*. Besides the above three important speeds of any power curve, there is one more speed which is specific together with the power curve: survival speed. It is that minimum speed at which the wind converter would not be able to sustain for its survival. The survival speed being more than cut-off speed, is not shown in the power curve. Its value is around 50–60 m/s. This value becomes one very important factor when selecting a wind turbine. It should be ensured that the maximum ever wind velocity in that location is lower than the survival speed of the machine.

It can be noted from the power curve that at lower wind speeds, the power output drops off sharply. This can be explained by the cubic power law, which states that the power available in the wind increases eight times for every doubling of wind speed and decreases eight times for every halving of the wind speed, as discussed in earlier chapters.

Using the power curve, it is possible to determine roughly how much power will be produced at the average or mean wind speed prevalent at a site. However, it is recommended to use the Weibull distribution for estimating the power output in connection with the power curve.

Power curves are calculated and found by field measurements, where an anemometer is placed on a mast reasonably close to the wind turbine (not on the turbine itself or too close to it, since the turbine rotor may create turbulence, and make wind speed measurement unreliable). If the wind speed is not fluctuating too rapidly, then one may use the wind speed measurements from the anemometer and read the electrical power output from the wind turbine and plot the two values together in a graph. In practice the wind speed always fluctuates, and one cannot measure exactly the column of wind that passes through the rotor of the turbine. It is not a workable solution just to place an anemometer in front of the turbine, since the turbine will also cast a "wind shadow" and brake the wind in front of itself.

In practice, therefore, one has to take an average of the different measurements for each wind speed, and plot the graph through these averages. Furthermore, it is difficult to make exact measurements of the wind speed itself. If one has a 3% error in wind speed measurement, then the energy in the wind may be 9% higher or lower (remember that the energy content varies with the third power of the wind speed). Consequently, there may be some errors in certified power curves. Nevertheless, a power curve is a very useful guide for estimating the output of a wind turbine.

6.1.2 Capacity Factor

Capacity factor is one element in measuring the productivity of a wind turbine or any other power production facility and to compare different locations with each other. It compares the plant's actual production over a given period of time, e.g. one year, with the amount of power the plant would have produced if it had run at full capacity for the same amount of time.

A conventional utility power plant, unless used as a peak load power plant, will normally run during most of the time unless it is idle only due to equipment problems or for maintenance or due to a reduced demand of energy. A capacity factor of 40% to 80% is typical for conventional plants. A wind plant is "fueled" by the wind, which does not blow steadily. Although modern utility-scale wind turbines typically operate 65% to 90% of the time, they often run at less than full capacity. Therefore, a capacity factor of 20% to 40% is common over one year, although they may achieve higher capacity factors during windy weeks or months and lower capacity factors in windless duration.

In Germany, another term, i.e. "load duration" is used to indicate the capacity factor. The term is a multiple of the total number of hours in one year (8,760 h) and the capacity factor. Since the capacity factor is much lower than one, the load duration also comes out to be much less than 8,760. For example, if the capacity factor is 20%, the load duration would be 8,760 × 0.2, which is 1,752 h.

The significance of load duration is that it expresses that number of hours for which the wind turbine can be considered to be virtually operating at its rated capacity in one year.

Here it is important to distinguish between plant availability factor and plant capacity factor. The first term refers to a fraction of one complete year for which the plant is available for use, irrespective of the availability of wind. For example, if a plant is under maintenance for 200 h in one year, out of a total of 8,760 h of one year, the availability factor would then be: (8760–200)/8760, i.e. 0.9771 or 97.71%.

To understand the capacity factor, let us consider a case of one 2 MW wind turbine:

In one year, a 2 MW wind turbine (8,760 h) can theoretically produce energy: $8,760 h \times 2\,MW = 17,520\,MWh$

If it actually produced 4,000 MWh only due to fluctuations in wind availability, the ratio of these two would be the capacity factor calculated below: Capacity factor $= 4000/17520 = 0.2283$ or 22.83%

It is important to note that while the capacity factor is almost entirely a matter of reliability for a fueled power plant, this is not the case with a wind plant. In case of a wind plant, it is a matter of economical turbine design. With a very large rotor and a very small generator, a wind turbine would run at full capacity whenever the wind blew and would have a 60–80% capacity factor, but it would produce very little electricity. The most electricity per unit of investment is gained by using a larger generator and accepting the fact that the capacity factor will be lower as a result. Wind turbines are fundamentally different from fueled power plants in this respect due to a limited and fluctuating availability of wind.

If a wind turbine's capacity factor is 33%, this does not mean it is only running one-third of the time. Rather, a wind turbine at a typical location would normally run for about 65–90% of the time. However, much of the time it will be generating at less than full capacity, making its capacity factor lower.

6.2 Power Control of Wind Turbines

Wind turbines are designed and operated to produce electrical energy as cheaply as possible. Wind turbines are, therefore, generally designed such that they yield a rated power output at wind speeds around 12–15 m/s. As discussed in previous sections, the exact value of such a wind velocity varies with manufacturer and size due to the difference in power curve. Its does not pay to design turbines that maximize their output at stronger winds, as such strong winds are rare. In case of stronger winds it is necessary to waste part of the excess energy of the wind in order to avoid damage to the wind turbine. Figure 6.2 shows the decline power coefficient for wind velocities exceeding the rated velocity.

Figure 6.2 shows the power curve of a 2.5 MW machine together with two more important curves: input power of wind together with the curve of multiplication of mechanical efficiency (η) and power coefficient (Cp). The product of these two represents the net efficiency of the wind turbine. The deviation between the input power and the power curve is due to net efficiency "$\eta \times c_p$". It can be very well observed that the curve for net efficiency reduces after reaching its maximum slightly before the rated wind speed. It is due to the fact that the pitch control starts wasting energy of wind after this point maintaining power under limits.

All wind turbines are designed with some sort of power control. There are two different ways of doing this safely on modern wind turbines.

6.2.1 Pitch Control

On a pitch controlled wind turbine the turbine's electronic controller checks the power output of the turbine several times per second. When the power output becomes too high, it sends an order to the blade pitch mechanism which immediately pitches (turns) the rotor blades slightly out of the wind. Conversely, the blades

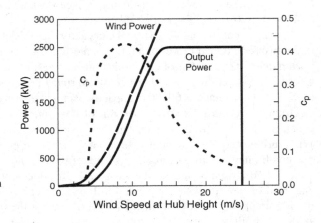

Fig. 6.2 Capacity utilization and power curve of a wind turbine

Fig. 6.3 Variable pitch
blades

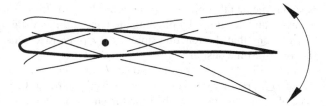

Fig. 6.4 Comparison of wind
turbines with fixed and
variable pitch control

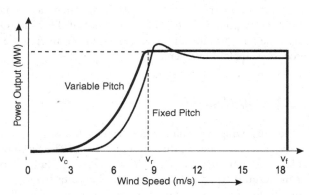

are turned back into the wind whenever the wind drops again. The rotor blades thus
have to be able to turn around their longitudinal axis (to pitch) as shown in Fig. 6.3.

During normal operation, the blades will pitch by fraction of a degree at a time -
and the rotor will be turning at the same time. The pitch mechanism can be operated
using hydraulic systems. But in most cases, individual electric drives are used to
actuate control of blades, and the same mechanism is also used for applying brakes
to the rotor by turning just one or two blades.

The importance of pitch control is clearly visible in Fig. 6.4. In the fixed pitch
machine, power output dips very quickly at a wind speed higher than the rated wind
speed and is oscillating around the rated power by higher wind speeds.. This can be
understood in connection with Fig. 3.5 in Chap. 3, showing the relationship between
power coefficient and tip speed ratio. As the wind speed increases, for a fixed rate
of revolution of the rotor, the tip speed ratio decreases. As a result, the power coefficient decreases very sharply, which reduces the power output from the wind mill
sharply as well. Whereas, in pitch controlled machines, due to a change of the blade
angle, the curve itself changes which avoids a sudden dip in the power coefficient
and output power.

6.2.1.1 Running a Pitch Controlled Turbine at Variable Speed

There are a number of advantages of being able to run a wind turbine at variable
speed. One good reason for wanting to be able to run a turbine partially at variable

speed is the fact that pitch-control (controlling the torque in order not to overload the gearbox and generator by pitching the wind turbine blades) is a mechanical process. This means that the reaction time for the pitch mechanism becomes an important factor in turbine design.

In a variable slip generator, however, one may start increasing its slip once it is close to the rated power of the turbine. The control strategy is to run the generator at half of its maximum slip when the turbine is operating near the rated power. When a wind gust occurs, the control mechanism signals to increase the generator slip to allow the rotor to run a bit faster, while the pitch mechanism begins to cope with the situation by pitching the blades more out of the wind. Once the pitch mechanism has done its work, the slip is decreased again. In case the wind suddenly drops, the process is applied in reverse.

6.2.2 Stall Control

An aircraft wing will stall, if the shape of the wing tapers off too quickly as the air moves along its general direction of motion. The turbulence is created on the back side of the wing in relation to the air current. Stall can be provoked if the surface of the aircraft wing – or the wind turbine rotor blade – is not completely even and smooth. A dent in the wing or rotor blade, or a piece like a self-adhesive tape can be enough to start the turbulence, even if the angle of the wing in relation to the general direction of airflow is small. Aircraft designers obviously try to avoid stall at all costs, since an aero plane without the lift from its wings will start falling down. Similarly, stall condition arriving in a wind energy converter would result in a no output condition.

In order to avoid stall condition, the rotor blades for large wind turbines are always twisted. Seen from the rotor blade, the wind will be coming from a much steeper angle (more from the general wind direction in the landscape), as one moves towards the root of the blade, and the centre of the rotor. As discussed above for stall, a rotor blade will stop giving lift, if the blade is hit at an angle of attack which is too steep. This is an additional reason for twisting blades to achieve an optimal angle of attack throughout the length of the blade. However, in the case of stall controlled wind turbines in particular, it is important that the blade is built such that it will stall gradually from the blade root and outwards at high wind speeds.

6.2.2.1 Passive Stall Control

Passive stall controlled wind turbines have the rotor blades bolted onto the hub at a fixed angle. The geometry of the rotor blade profile, however, has been aerodynamically designed to ensure that at the moment the wind speed becomes too high, it creates turbulence on the side of the rotor blade which is not facing the wind. This stall prevents the lifting force of the rotor blade from acting on the rotor.

The basic advantage of stall control over pitch control is that one avoids moving parts in the rotor itself, and a complex control system. On the other hand, stall control represents a very complex aerodynamic design problem, and related design challenges in the structural dynamics of the whole wind turbine, e.g. to avoid stall-induced vibrations. A normal passive stall controlled wind turbine will usually have a drop in the electrical power output for higher wind speeds, as the rotor blades go into deeper stall. Therefore, large wind machines are not using stall control for controlling output power.

6.2.2.2 Active Stall Control

An increasing number of larger wind turbines (1 MW and more) are being developed with an active stall power control mechanism. Technically the active stall machines resemble pitch controlled machines, since they have pitchable blades. In order to get a reasonably large torque (turning force) at low wind speeds, the machines will usually be programmed to pitch their blades much like a pitch controlled machine at low wind speeds. (Often they use only a few fixed steps depending upon the wind speed). When the machine reaches its rated power, however, an important difference from the pitch controlled machines can be noticed. The difference is that if the generator is about to be overloaded, the machine will pitch its blades in the opposite direction from what a pitch controlled machine does. In other words, it will increase the angle of attack of the rotor blades in order to make the blades go into a deeper stall, thus wasting the excess energy in the wind.

One of the advantages of active stall is that one can control the power output more accurately than with passive stall, so as to avoid overshooting the rated power of the machine at the beginning of a gust of wind. Another advantage is that the machine can be run almost exactly at rated power at all high wind speeds.

6.2.3 The Yaw Control

Some older wind turbines use flaps to control the power of the rotor, just like air-crafts use flaps to alter the geometry of the wings to provide extra lift at the time of takeoff. Another theoretical possibility is to yaw the rotor partly out of the wind to decrease power. This technique of yaw control is in practice used only for tiny wind turbines, as it subjects the rotor to cyclically varying stress which may ultimately damage the entire structure. The wind turbine yaw mechanism is used to turn the wind turbine rotor against the wind.

The wind turbine is said to have a yaw error, if the rotor is not perpendicular to the wind. A yaw error implies that a lower share of the energy in the wind will be running through the rotor area. The share will drop to the cosine of the yaw error, due to the change in the projected area of the rotor facing the wind.

If this were the only thing that happened, then yaw control would be an excellent way of controlling the power input to the wind turbine rotor. The main problem with yaw control is that the part of the rotor which is closest to the source direction of the wind, however, will be subject to a larger force (bending torque) than the rest of the rotor. On the one hand, this means that the rotor will have a tendency to yaw against the wind automatically, regardless of whether we are dealing with an upwind or a downwind turbine. On the other hand, it means that the blades will be bending back and forth in a flapwise direction for each turn of the rotor. Wind turbines which are running with a yaw error are therefore subject to larger fatigue loads than wind turbines which are yawed in a perpendicular direction against the wind.

6.3 Connection to the Grid

The connection to grid depends upon the purpose for which a wind energy system is used. Although most large wind energy systems are installed to feed electricity to the grid, however, other configurations are also found, particularly in small and medium size turbines. Their requirements for grid connection are described in this section.

6.3.1 Applications of Wind Energy Converters

From the point of view of utilization of generated electricity, wind energy systems can be used in three ways as explained below:

(a) *Grid connected system.* The grid connected system has a connection with an electricity transmission and distribution system called grid-connected systems. A grid-connected wind turbine can be used for reducing the consumption of fossil energy carriers. If the turbine cannot deliver the amount of energy needed, the public utility makes up the difference.

(b) *Stand alone system.* As their name suggests, stand-alone systems are not connected to the utility grid. Stand-alone wind energy systems can be appropriate for homes, farms, or even entire communities (a co-housing project, for example) that are far from the nearest utility lines. However, if grid is available in a nearby area, it is always advantageous to be connected to the grid. Therefore, stand alone systems are used only in areas where grid is not available at all.

(c) *Hybrid systems.* A hybrid system using wind energy is one in which operation of a wind turbine is combined with any other source of power. Such a combination can be considered as another type of stand alone system. This source may be e.g. a photovoltaic system, or a diesel generator set. To specify the nature of these systems, sometimes the term "stand-alone hybrid system" is also used to differentiate between the hybrid systems in which wind turbine is operating in hybrid mode with the grid supply.

At many sites, wind speeds are low in the summer when the sun shines brightest and longest. The wind is strong in the winter when there is less sunlight available. Since the peak operating times for wind and photovoltaic systems occur at different times of the day and year, hybrid systems are more likely to produce power when required.

For the times when neither the wind generator nor the photovoltaic modules are producing electricity (for example, at night when the wind is not blowing), most stand-alone systems provide power through batteries and/or an engine-generator powered by fossil fuels like diesel.

If the batteries run low, the engine-generator can be run at full power until the batteries are charged. Adding a fossil-fuel-powered generator makes the system more complex, but modern electronic controllers can operate these complex systems automatically. Adding an engine-generator can also reduce the number of photovoltaic modules and batteries in the system. It should be ensured that the storage capability is large enough to supply electrical needs during non-charging periods. Battery banks are typically sized for one to three days of windless operation.

6.3.2 Voltage Requirement

A single small converter can be directly connected into the grid network at e.g. 0.4 kV level. Once the wind energy converter is integrated into the grid network, there must be very limited voltage change, voltage oscillation or flicker experienced in the homes on that network branch. The loss of voltage due to resistance in the cabling can be avoided by increasing the diameter of the cables. It is often required that a new network branch is constructed and linked to the transformer in order to reduce the voltage disturbances. This increases the installation costs of the converter.

A single Mega-Watt size converter cannot be connected to the grid at the 0.4 kV stage, but has to be connected at 10–30 kV, which is the usual level of the city electricity share distribution. In remote areas, where a 30 kV connection is not established, the connection must be created and financed. Wind parks with a lot of Mega-Watt converters must be connected into the electrical grid at a level of about 100 kV and higher.

As mentioned earlier, the maximum power output is obtained only in a few hours during the year. Figure 6.5 shows a typical load distribution, measured within the German 250 MW program. With larger wind energy installations, this uneven distribution leads to the need of higher regulation capacities by conventional power systems in the future.

It would be worth mentioning here that due to fluctuations in availability of wind, these systems alone cannot fully meet the energy demand. On a large scale, additional conventional power plants that are flexible in their level of operation are needed. Whenever wind is available, the level of operation of the conventional power plant is reduced, and when wind is not available, it is increased to meet the demand.

6.3.3 Special Aspects of the Connection of Offshore Wind Parks

Grid connection of offshore wind farms is a technical and economical challenge to both wind turbine and grid operators. In the initial phase, the still quite limited capacity of early pilot farms allows the use of a conventional three-phase AC

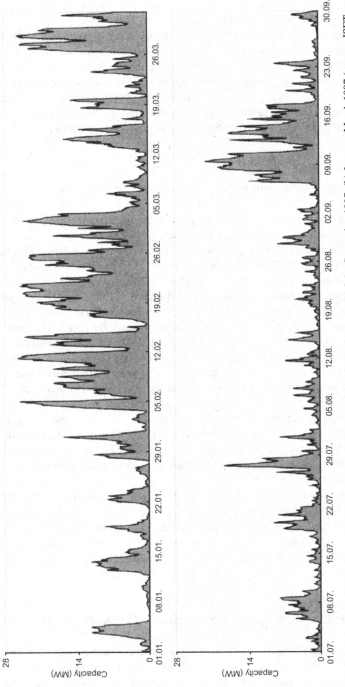

Fig. 6.5 Load Distribution of a wind park with a total capacity of 28 MW in Germany (**a**) July–September 1997. (**b**) January–March 1997 (source: ISET, see literature)

connection to the onshore grid system which is a well known and inexpensive technology.

An internal grid is necessary to connect the offshore wind farm to the onshore grid. The produced power has to be fed to an offshore transformer substation, to which wind turbines are connected via undersea cables by a voltage of e.g. 30 kV. After stepping-up to the transmission line voltage, the power is conveyed to the shore.

Better suitability of submarine High Voltage DC transmission lines for offshore wind turbines is still a debatable issue. High Voltage DC lines offer the advantage of reduced transmission losses but they have the problem of higher cost. The issue is much more important as compared to on-shore turbines due to greater lengths of transmission, which are of the order of 5–10 km.

Chapter 7
Economics and Policy Issues

This chapter presents various economic and financial considerations that are important for selecting wind turbines, and before taking a decision in favor of installing a
wind turbine at all.

7.1 Cost of Wind Turbines

7.1.1 Initial Cost of Wind Turbine

The average price of large, modern onshore wind farms in Germany is around 1,250
€ per kilowatt electrical power installed, in India it is slightly less than 1,000 €
(65,000 Indian Rupees) per kilowatt, mainly due to cheaper labor costs. These costs
are just indicative figures and would differ with capacities and make. For single
turbines or small clusters of turbines the costs will usually be somewhat higher as
compared to large wind farms, due to bulk purchase price breaks. These costs also
increase with an increase in tower height.

Installation costs include foundations, road construction (necessary to move the
turbine and the sections of the tower to the building site), a transformer (necessary to
convert the low voltage e.g. 690 V) current from the turbine to 10–30 kV current for
the local electrical grid, telephone connection for remote control and surveillance of
the turbine, and cabling costs, i.e. the cable from the turbine to the local 10–30 kV
power line.

Obviously, the costs of roads and foundations depend on soil conditions, i.e.
how cheap and easy it is to build a road capable of carrying heavy trucks. Another
variable factor is the distance to the nearest ordinary road, the cost of getting a
mobile crane to the site, and the distance to a power line capable of handling the
maximum energy output from the turbine. Transportation costs for the turbine may
enter the calculation, if the site is very remote. Of course this figure also depends
upon the weight of the turbine, distance to be covered and, last but not least, current
fuel prices.

It is obviously cheaper to connect many turbines at the same location, than just
one. On the other hand, there are limits to the amount of electrical energy the local

H.-J. Wagner, J. Mathur, *Introduction to Wind Energy Systems*, Green Energy and
Technology, DOI 10.1007/978-3-642-02023-0_7, © Springer-Verlag Berlin Heidelberg 2009

electrical grid can handle. Especially in developing countries like India, if the local grid is too weak to handle the output from the turbine, a need for grid reinforcement may arise, i.e. by extending the high voltage electrical grid. Agency that bears the cost of grid reinforcement: the power company or the turbine owner; varies from country to country.

7.1.2 Operation and Maintenance Costs for Wind Turbines

Modern wind turbines are designed to work for some 120,000 h of operation throughout their design lifetime of 20 years.

7.1.2.1 Operation Maintenance Costs

Experience shows that maintenance costs are generally very low when the turbines are brand new, but they increase somewhat as the turbines age. The newer generations of turbines have relatively lower repair and maintenance costs than the older generations. Older wind turbines (25–150 kW) require annual maintenance costs at an average of around 3% of the original turbine investment. Newer turbines are on average substantially larger, which would tend to lower maintenance costs per kW installed power, since it is not needed to service a large, modern machine more often than a small one. For newer machines the estimates range around 1.5–2% per year of the original turbine investment. In some cases, however, e.g. if the product is not standardized and/or the site is not investigated thoroughly, even these costs may be relatively high as compared to industry average.

Most of the maintenances cost are expressed as a fixed amount per year for the regular service of the turbines, but some people prefer to use a fixed amount per kWh of output in their calculations, for example around 0.01 €/kWh. The reasoning behind this method is that tear and wear on the turbine generally increases with increasing production.

In addition to maintenance, another important cost element is insurance costs. Insurance of wind turbines is required to secure the loss of their heavy investment (fully or partially as damage of parts) due to any unforeseen damage, such as lightning or hurricanes, or any other factor.

Other than the economies of scale which vary with the size of the turbine, as mentioned above, there may be economies of scale in the operation of wind parks rather than individual turbines. These economies are related to the semi-annual maintenance visits, surveillance and administration, etc.

7.1.2.2 Turbine Reinvestment (Refurbishment, Major Overhauls)

Some wind turbine components are more subject to tear and wear than others. This is particularly true for rotor blades and gearboxes. Wind turbine owners who see that their turbine is close the end of their technical design lifetime may find it advantageous to increase the lifetime of the turbine by doing a major overhaul of the turbine, e.g. by replacing the rotor blades after 7–10 years. The price of a new set of rotor blades, a gearbox, or a generator is usually in the order of magnitude of 15–20% of the price of the turbine.

7.1.2.3 The Availability Factor

The figures for annual energy output assume that wind turbines are operational and ready to run all the time. In practice, however, wind turbines need servicing and regular inspection (e.g. once every six months) to ensure that they remain safe. In addition, component failures and accidents (such as lightning strikes) may disable wind turbines. The best turbine manufacturers consistently achieve availability factors up to 98%, i.e. the machines offer technical availability of up to 98% of the time. Total energy output is generally affected by less than 2%, since wind turbines are never serviced during high winds. The availability factor is, therefore, usually ignored when doing economic calculations, since other uncertainties (e.g. wind variability) are far larger.

It is always a good idea to check the manufacturers' track record and servicing ability before buying a new wind turbine. Manufacturers in most cases offer a guarantee of availability of the machine for the initial 2–3 years. Afterwards, they prefer to enter into a sort of annual maintenance contract with the machine owner, in which they charge a fixed amount to guarantee a certain minimum availability of the system for the generation of power.

7.2 Electrical Tariffs

Five different types of tariff systems are found in connection with wind energy systems due to a difference in policies which change in country/location as described below:

(a) *Feed-in tariff.* The feed-in tariff scheme, as its name suggests, is based upon the principle of paying an amount to the wind energy producer that depends on the amount of electricity fed into the grid. This is done at a pre-declared rate per unit of electricity. This rate is higher than the rate of production of electricity from a conventional (using fossil or nuclear fuels) power plant. The most important aspect in a feed-in tariff system is that the grid cannot deny accepting the power generated by the wind energy system, even if it is surplus. The feed-in tariff system exists e.g. in Germany as well as in many other countries. The rates of feed-in tariff change with respect to the location within the country. To promote the use of wind energy systems by improving their economics, currently, in Germany, the feed-in-tariff rates are higher in inland locations as compared to coastal locations which are attractive for wind energy anyway.

If an energy producer owning wind turbines produces a certain amount of electricity and consumes the same amount of electricity, separate accounting is done for the electricity produced by the windmill and the electricity consumed. The feed-in tariff rate is not uniform for all renewable energy systems, e.g. in the case of photovoltaic systems, the feed-in tariff rates are even higher than those of wind energy systems.

Feed-in tariff rates also exist in India, but a difference in feed-in tariff rates between coastal and non-coastal areas does not exist. Currently, the rates are the same over one particular state; they are the highest in the state of Maharashtra and the lowest in the state of Tamilnadu.

(b) *Net metering system*. Net metering or net billing is a term applied to laws and programs under which a utility allows the meter of a customer with a wind turbine, to turn backward, thereby, in effect, allowing the customer to deliver any excess electricity he produces to the utility and be credited on a one-for-one basis against any electricity the utility supplies to him. For example: During a one-month period, one wind turbine in a farm house generates 300 kilowatt-hours (kWh) of electricity. Most of the electricity is generated at a time when the equipment in the farm house (refrigerator, lights, etc.) draws electricity, and the power generated is used on site. However, some is generated at night when most equipment is turned off. At the end of the month, the turbine has generated 100 kWh in excess of local needs, and that electricity has been transmitted to the utility system. Over the same one-month period, the utility system supplied to the farm house electricity of 500 kWh for use at times when the wind turbine was not or insufficiently generating. Since the meter ran backward while the 100 kWh was being transmitted to the utility system, the farm house will be billed for 400 kWh rather than 500 kWh.

The net-metering scheme can improve the economics of captive wind turbine by allowing the turbine's owner to use the excess electricity to offset utility-supplied power at the full retail rate, rather than having to sell the power to the utility at the price the utility pays for the wholesale electricity it buys or generates itself. Many utilities argue against net metering laws, saying that they are being required, in effect, to buy power from wind turbine owners at full retail rates, and are, therefore, being deprived of a profit from part of their electricity sales. However, wind energy advocates have successfully argued that what is going on is a power swap, and that it is standard practice in the utility industry for utilities to trade power among themselves without accounting for differences in the costs of generating the various kilowatt-hours involved.

Although in Europe the feed-in tariff scheme exists, the popularity of net metering is growing, worldwide. Some countries like India even offer limited flexibility in the installation of wind turbines and consumption centers. This means that they may not be exactly at the same place. In such cases, of course, the utility charges a nominal amount termed as "wheeling charges" against use of their infrastructure/power lines for transferring power from the point of generation to the point of consumption. In such arrangements, there are, of course, some limitations such as geographical and commercial boundaries which differ from state to state.

(c) *Time dependent rates*. Ideally, electricity companies are more interested in buying electricity during the periods of peak load (maximum consumption) on the electrical grid. Therefore, in some areas, power companies apply variable electricity tariffs depending on the time of day, when they buy electrical energy from private wind turbine owners. Normally, wind turbine owners receive less than the normal consumer price of electricity, since that price usually includes payment for the power company's operation and maintenance of the electrical grid, plus its profits. In locations where a feed-in tariff system exists, as in Germany, the time when power is generated becomes non-significant.

(d) *Quota system.* An additional system is that of the allocation of "quota" of renewable energy. In this system, every producer of electricity for grid is given a "quota", e.g. 20%. In the total electricity produced by every company, there has to be 20% share of energy coming from renewable sources such as wind and solar. In case this quota is not met, provisions of penalties are made. The European Union is discussing to adopt this system in Europe. Provisions are being developed that if a company produces more energy from renewables than its quota, e.g. 25%, it would be granted a certificate for this excess renewable energy. This certificate can be purchased by other companies that are having a lower share of renewables from the quota, e.g. 15%. By purchasing these certificates, the second company will also be considered to comply with its quota. This system is quite similar to the concept of "emission trading" for greenhouse gases. Up to the middle of 2008, there was no consensus on bringing such a quota system to Europe. The argument was debated that with such a quota system, relatively costlier renewable energy technologies, such as solar photovoltaic systems, would not be installed at all. Companies might prefer to purchase certificates in place of installing PV systems. The concept of the quota system is still being discussed at EU level.

(e) *Production tax incentives/Investment incentives.* The Production tax incentive is a generation-based mechanism, which supports renewable energies through payment exemptions from electricity taxes, e.g. the energy tax for renewable energies, applied to all producers. Hence it is a system that affords an avoided cost on the producer side. Also the Investment incentive is a mechanism, like the name leads one to divine, to lower the costs for the investment in renewable energies so that it gets more attractive funding.

7.3 Mechanisms to Support Funding

7.3.1 Capacity Credit

To understand the concept of the capacity credit, we look at its opposite, the power tariffs: Large electricity customers are usually charged both for the amount of energy (kWh) they use, and for the maximum amount of power (kW) they draw from the grid, i.e. customers having more connected load drawing a lot of energy very quickly have to pay more. The reason they have to pay more is that this fact obliges the power company to have a higher total generating capacity (more power plant) available. Power companies have to consider adding generating capacity whenever they give new consumers access to the grid. But with a modest number of wind turbines in the grid, wind turbines are almost like "negative consumers", since they postpone the need to install other new generating capacity. Many power companies in industrialized countries, therefore, pay a certain amount per year to the wind turbine owner as a capacity credit. The exact level of the capacity credit varies with location benefit to utility in terms of avoided installation of power plants. In some countries it is paid on the basis of a number of measurements of power output during the year. In other areas, some other formula is used. In a number of countries like Germany, no capacity credit is given, as it is assumed to be part of the energy tariff due to the fact that due to fluctuating output from wind energy system, the utility anyways has to install power plants as back up for a no wind situation.

7.3.2 Environmental Credit and Clear Development Mechanism

Many governments and power companies around the world wish to promote the use of renewable energy sources. Therefore, in developed or industrial countries, they offer a certain environmental premium to wind energy, e.g. in the form of a refund of electricity taxes etc. on top of normal rates paid for electricity delivered to the grid. In developing countries like India, due to the fact that every unit of electricity generated avoids generation of the same amount of electricity from fossil fuel based power plants, the avoided environmental emissions offer the opportunity for additional earnings through the Clean Development Mechanism (CDM) under the Kyoto Protocol. Every ton of carbondioxide gas saved or avoided, termed as one carbon credit, is sold to industrialized countries at rates of around 15 €. These additional earnings help to improve the economics of power generation, which has been one strong reason behind the rapid increase in the wind energy based installed capacity for power generation in India.

7.3.3 Tax Benefits

In some countries like India, several tax benefits are offered to promote wind energy systems. Major incentives or benefits are mentioned here:

(a) If the wind energy converter or its parts or components are imported from other countries, no excise duty (import tax) is applicable.
(b) Exemption/reduction in central government sales tax and general sales tax are available on the sale of renewable energy equipment in various states of India.
(c) 80% accelerated depreciation on specified renewable energy devices/systems (including wind power equipment) in the first few years of installation of the projects.
(d) There is an income tax holiday on the income generated from renewable energy systems including wind energy through power generation.

As a result of these measures, several investors have installed wind energy systems in different parts of India.

7.4 Wind Energy Economics

7.4.1 Financial Analysis – A Case Study for India

The annual electricity production will vary enormously depending on the amount of wind on the turbine site. Therefore, there is not a single price for wind energy. It differs from year to year and location to location, that means wind speed and wind profiles.

Following a case study from India should give an impression about the factors which has to be considered to calculate the electricity costs from wind turbines.

It is proposed to install a 1,250 kW wind power plant at a certain location where the capacity utilization for power generation over one year is expected to be 35%. Below are the major cost elements in Indian rupees (Rs.):

Currency: Rs. 100 = 1,54 €; 1 € = Rs. 65
Cost of land: Rs. 1 million
Cost of wind energy converter: Rs. 60 million
Erection and commissioning charges: Rs. 6 million
Sub station charges: Rs. 4.0 million

Operation and maintenance charges: Free during first year, second year Rs. 1.1 million, for subsequent years with an escalation of 5% per year.

Other economic parameters include a total generation as 3.85 million kWh electricity (at 35% capacity utilization), and losses as 7%.

Tariff for selling the electricity is Rs. 5.5/kWh in the first year, with a price escalation of 2% per year.

Wheeling charges: 5% of tariff

Life cycle term of the project is 20 years.

The investment requirement, power generation and sale of power can be summarized as in the Table below (Table 7.1):

Table 7.2 shows a detailed cash flow analysis of the project under various heads:

It can be seen from the above table that the cumulative cash accrual becomes equal to the initial investment of Rs. 71 million at the end of the fourth year. This period is known as the simple payback period of investment. Beyond this period and up to the end of project life (20 years in this case), every cash accrual is net profit from the project. If the financial discounting rate is significant and the payback period is longer, it is recommended to include the time based value of money for estimating the payback period.

7.4.2 Financing of a Wind Park – A Case in Germany

Due to significant differences in the amounts invested for an onshore and off-shore wind park, there are two different cases to describe. The amount invested

Table 7.1 Assumptions for the case study of financial analysis

Investment requirement	
Cost of land (million Rs.)	1
Cost of wind machine (million Rs.)	60
Erection and commissioning charges (million Rs.)	6
Sub-station charges (million Rs.)	4
Total cost (million Rs.)	71
Running cost requirement	
O & M (million Rs.) per year	1.1
Escalation	5%
Free O & M (Years)	1
Depreciation rate	80%
Insurance (million Rs./year)	0.15
Tariff (Rs./kWh)	5.5
Escalation in tariff per year	2%
Wheeling charges (% of tariff)	5%
Generation (million kWh)	3.85
Losses	7%
Sale of power	
Tariff (Rs./kWh)	5.5
Wheeling charges (5% of tariff)	0.28
Actual tariff (tariff – wheeling charges) (Rs./kWh)	5.23
Generation (million kWh/year)	3.85
Losses(7% of generation)	0.27
Actual generation (generation – losses) (million kWh/year)	3.58

for an e.g. 15 MW onshore wind park is about 19 million €, which is split in an equity capital of at least 4 million € and a dept capital of 15 million €. The wind power plant, the installation, the commissioning and the infrastructure constitutes

Table 7.2 Cash flow analysis of a wind energy project

Year	1	2	3	4	5	6–20
Receipt (million Rs.)						
Power rate (Rs./kWh)	5.23	5.28	5.33	5.38	5.44	Increasing as per rate of tariff
Receipt (sale of power)	18.71	18.90	19.08	19.27	19.47	Increasing from 6th year to 20th year
Total receipts (A)	18.71	18.90	19.08	19.27	19.47	Increasing from 6th year to 20th year
Expenses (million Rs.)						
O & M (free for 1st year)	0	1.1	1.155	1.21	1.27	Increasing from 6th year to 20th year
Insurance	0.15	0.15	0.15	0.15	0.15	0.15
Total expenses (B)	0.15	1.25	1.31	1.36	1.42	Increasing from 6th year to 20th year
Cumulative cash accrual (million Rs.) (A–B)	18.56	36.20	53.98	71.89	89.94	Increasing up to 20th year (project life)

an 83% proportion of the investment volume. With 11%, technical planning, surface investigation, a survey report as well as the grid connection costs also capture
a capacious amount of the investment volume. Compensatory measures, compen-
sations for use during the building-up phase, cost for the partnership during the
investment phase, the consulting and the financing during the construction period
adopt rather inferior parts.

To raise the equity capital a limited partnership (L.P.) is founded. This L.P. incurs
the sale of the interests and covers the banker's guarantee. It is also responsible for
the construction, the instigation and the disposal of the produced electric power.
The limited partner's individual deposit constitutes the equity capital and starts from
several thousand Euros per partner. The bank allocates the dept capital in form of
a loan. This loan could be for example a loan of 11 million €, with a duration of
15 years and an interest loan of about 6%. As a chattel mortgage the wind park, the
feed-in tariffs and the liquidity reserves are transferred to the bank until the loan is
paid off. The money for the bank has priority over the payout of dividends for the
limited partners.

The funding of an offshore wind park requires an investment volume of at least
200 million € and more. Therefore, a limited partnership cannot be established and
grand transmission providers have to take over this part. Due to the lack of experience of the instigation of offshore parks up to now it is difficult to find cooperative
banks to finance these parks, which retard the process of the installation of win
parks.

In such a case it could be helpful if the government stand security for the
financing of the first windparks.

7.5 Wind Turbines After Operational Life

The amount of material used in wind turbines increase with the size and the number
of wind turbines. After the operational life of the equipment it must be demolated
and as far as possible recycled. Recycling of steel and other metals do not create
problems, whereas the rotor blades, made of plastic, glass fibers or carbon fibers are
at least difficult to recycle. Therefore industry should develop solutions to get rid of
the waste in ecological acceptable waste treatment produces.

Chapter 8
Outlook

More than 50 countries around the world have an organized set up in the field of wind power. Large progress in wind power has been witnessed in the countries of the European Union, but the situation is changing quite rapidly. The United States and Canada are both experiencing a surge of activity, while markets like Asia and South America are also catching up quite rapidly. In the past few years, a new frontier for wind power development has also been established in the sea. With offshore wind parks beginning to make a contribution, the possibilities of using wind energy systems worldwide are likely to increase many times in the future. Establishing wind energy projects in the sea has opened up new demands, including the need for stronger foundations, long underwater cables and larger individual turbines, but offshore wind parks are expected to contribute an increasing proportion of global capacity.

With its increase in the global market, wind power has seen a fall in cost. A modern wind turbine produces more electricity at less cost per unit (kWh) than its equivalent 20 years ago. At good locations wind may compete with, and even beat, the cost of both coal and gas-fired power, if the future carbon dioxide certificate costs are taken into account. If the "external costs" associated with the pollution and health effects resulting from fossil fuel and nuclear generation are fully taken into account, wind power gains an additional advantage.

Studies of different institutions around the world have confirmed that a lack of wind is not likely to be a limiting factor on global wind power development. With the expansion of the wind industry, large quantities of wind powered electricity will need to be integrated into the grid network. Because of the variability of the wind, control methods have to be established for dealing with variations in demand and supply. Their installation could help to handle this issue.

It has been estimated by the Global Wind Energy Council that wind power could supply one third of the world's electricity by 2050. This is a big challenge not only for the wind energy sector, but for the entire electricity supply system. Let's try to catch it.

H.-J. Wagner, J. Mathur, *Introduction to Wind Energy Systems*, Green Energy and
Technology, DOI 10.1007/978-3-642-02023-0_8, © Springer-Verlag Berlin Heidelberg 2009

Glossary

Alternating Current (AC): An electric current that reverses its direction at regularly recurring intervals, usually 50 or 60 times per second. Today's grids are operated by Alternating Current.

Direct Current (DC): An electric current with constant direction. It can be used on high voltage level to transport bulk electricity with less losses.

Gigawatt (GW): A unit of power equal to 1 billion watts, 1 million kilowatts, or 1,000 megawatts.

Joule (J): A standard international unit of energy; 1,055 joules are equal to 1 BTU.

Kilowatt (kW): A standard unit of electrical power equal to 1,000 watts, or to the energy consumption at a rate of 1,000 joules per second.

Kilowatt-Hour (kWh): 1,000 watts acting over a period of 1 hour. The kWh is a unit of energy. 1 kWh = 3,600 kJ.

Megawatt (MW): 1,000 kilowatts, or 1 million watts, standard measure of electric power plant generating capacity.

Megawatt-Hour (MWh): 1,000 kilowatts-hours or 1 million watt-hours.

Watt (W): The rate of energy transfer equivalent to one ampere under an electrical pressure of one volt. One watt equals 1/746 horsepower, or one joule per second. It is the product of voltage and current (Amperage).

Watt-hour (Wh): The electrical energy unit of measure equal to 1 watt of power supplied to, or taken from, an electric circuit steadily for 1 hour.

H.-J. Wagner, J. Mathur, *Introduction to Wind Energy Systems*, Green Energy and Technology, DOI 10.1007/978-3-642-02023-0_BM2, © Springer-Verlag Berlin Heidelberg 2009

Literature

Annual report 2007–2008. Centre of Wind Energy Technology, Government of India, http://cwet.tn.nic.in/

Annual report 2007–2008. Indian Renewable Energy Development Corporation, http://www.ireda.in/incentives.asp

Annual report 2007–2008. Ministry of New and Renewable Energy sources, India, http://www.mnes.ni.in/

Bansal, N K, Kleemann, M, Meliss, M (1990) Renewable Energy Sources and Conversion Technology. Tata McGraw-Hill Publishing Company Ltd., New Delhi, ISBN 0-07-4600023-0

Global Wind 2008 Report. Global Wind Energy Council http://gwec.net

Heier, S (2006) Grid integration of wind energy conversion systems. John Wiley & Sons Ltd., ISBN 978-0470868997

Landolt-Börnstein (2006) Energy Technologies, Subvolume C Renewable Energy. Springer Verlag S. 233–241, ISBN 3-540-42962-X

Pantel Mukund, R (2006) Wind and Solar Power Systems – Design, Analysis and Operation. CRC Press, ISBN 0849315700

Scientific Measure and Evaluation Program for the 250 MW-Wind Test Program, Results of 1997 (1998). Wissenschaftliches Mess- und Evaluierungsprogramm (WMEP) zum Breitentest 250 MW Wind, Jahresauswertung im Auftrag des Bundesministeriums für Bildung, Wissenschaft, Forschung und Technologie, Kassel: Institut für Solare Energieversorgungstechnike n (ISET) (in German language)

Index

UNCED process, and the crisis and the erosion of the Green movement itself.

Overall, the set-up of the UNCED Process clearly favoured the most powerful lobbyists. The first step of this set-up was the very definition of an 'NGO' or an 'independent sector', as the Center for Our Common Future and the IFC liked to call them. As defined by the UN, the acronym 'NGO' is a catch-all that covers anything that is not governmental: But while this encompasses anything from the best known activists like Greenpeace to religious groups like Hare Krishna, it also covers non-profit business associations whose real mission is to try to promote the sale of many of the things that activists are opposed to, from toxic chemicals to nuclear weapons. By uniformly referring to the various groups involved in UNCED as 'non-governmental' or 'independent', one is led to believe that they are all equally legitimate agents. Of course, the origin of this model stems from national politics, replicating the ideal of US or Canadian 'democracy', where all groups that can organize themselves have the theoretical chance to lobby and thus to influence government policy. The problem with this model is that the global political system is not set up like the US government. In the absence of a coherent government to be lobbied at the global level, the strongest – i.e. most powerful and financially most potent lobbyists – quickly substitute themselves for all other international agents. And this is exactly what happened. As we show in the next section, the Business Council for Sustainable Development or the International Chamber of Commerce took over even the very way environment and development problems were to be looked at.

This lobbying model furthermore implies that, in order to be an efficient lobbyist, NGOs and independent sectors have to organize. The ones which will get most out of UNCED will be those 'NGO-coalitions' or independent sectors which 'speak with one voice', as Maurice Strong once suggested to them. The environmental movement should of course have seen that this process of organizing in order to speak with one voice was ultimately going to weaken it. On the other hand, business and industry had understood that this lobbying model was offering them a unique competitive advantage. Unlike most other NGOs, business and industry already had a global presence and some – for example, the International Chamber of Commerce – a global organizational structure. Also, the mission of business and industry is much more unified and coherent than the missions of the various environmental and developmental NGOs. Indeed, a coherent mission makes it easier to organize.

BUSINESS GETS ORGANIZED FOR UNCED

Business and industry have perfectly conformed to the expectations of the lobbying model set up by Strong and the UNCED secretariat. Very rapidly, therefore, these sectors gained a considerable comparative advantage over all others. Indeed, they seemed to have heard the Brundtland Commission's call for sustainable development before all other independent sectors. In retrospect, one may ask whether the business and industry sectors did not receive some insider information, or at least friendly suggestions.

The Brundtland Commission had hardly started when, in 1984, the International Chamber of Commerce (ICC), in collaboration with UNEP, organized the first World Industry Conference on Environmental Management (WICEM I) in Versailles, France. The outcome was the emergence, in 1986, of the International Environmental Bureau (IEB), which was originally located with the World Economic Forum in Geneva and is now with the ICC office on environment and energy in Norway. IEB was, at that time, a trans-industry clearinghouse on environmental management information. The Bergen conference was the next significant step in the business sector's endeavours aimed at Rio: out of the Bergen conference and the parallel Industry Forum came the 'European Green Table', 'a contribution to the work of the ICC towards the 1992 UNCED'.[2] In Bergen ICC was mandated to prepare seven industry projects, to form the core of an industry initiative, to be finalized at WICEM II. 'This initiative will both cover industry's own operational approach to sustainable development, and prepare the main policy issues relevant to world industry in relation to UNCED'.[3]

WICEM II, the second World Industry Conference on Environmental Management, was held in Rotterdam in April 1991. At that occasion, *Network '92* (as the journal of the Center for Our Common Future was then titled), generally reflecting the views of the global environmental managers, remarked: 'As the first global sectorial initiative organized to prepare for the Earth Summit, the organizers of WICEM are to be congratulated on being first off the mark and for a well structured effort'.[4] In retrospect, this remark sounds quite cynical.

In the mean time, Maurice Strong had appointed Dr Stephan Schmidheiny — 'a leading Swiss industrialist'[5] — as the principal adviser for UNCED. Dr

Schmidheiny recruited a group of forty-eight business leaders from around the world and during WICEM II created the Business Council for Sustainable Development (BCSD). Said the *Brundtland Bulletin* at that time:

The Business Council will provide advice and guidance to the UNCED secretariat on initiatives and activities undertaken by business and industry in respect of the preparatory process for the 1992 Conference, including programmes developed by the International Chamber of Commerce and other business organizations and bodies, programmes developed by the World Economic Forum and its industry Fora, and programmes developed by individual corporations and business leaders.[6]

Interestingly, 'Maurice Strong requested that the mandate be carried out well in advance of the Earth Summit so that the input of the Business Council's members could be taken into consideration during the consultative process that the UNCED Secretary General is carrying out prior to Rio'.[7] As a result, the BCSD fed directly into the 'consultative process' of UNCED, whereas most NGOs fed, if at all, into the discussions that went on at the Preparatory Committee's meetings.

Even though, as we see in the next section, the BCSD was very successful during and after the Earth Summit, differences seem to have emerged between Dr Schmidheiny on the one hand and the International Chamber of Commerce on the other. As a result, in February 1992 the ICC set up a new organization – the World Industry Council for the Environment (WICE) – 'to lobby on environmental issues for business interests'.[8] Over sixty international companies are founding members of WICE, i.e. more than in the BCSD. Nevertheless, before and during the UNCED process it was the BCSD that lobbied on behalf of business and industry.

THE 'SUSTAINABLE COUNCIL FOR BUSINESS DEVELOPMENT'

As we have said, the key figure in the BCSD is the Swiss billionaire Stephan Schmidheiny. The official story goes that his association with the Summit began in mid-1990 when Maurice Strong appointed him to be his principal adviser for business and industry. Their personal relationship, however, goes back to the Davos Forum – of which Maurice Strong had been the chairman – a

glamorous annual meeting of international business and government leaders in Switzerland, whose aim is to promote business–government dialogue. Schmidheiny's family owns Swatch and made a lot of money from investment in the asbestos industry, although after he started to work for the Summit he sold his asbestos holdings. He sits on the board of at least two TNCs: Asea Brown Boveri, manufacturers, among others, of nuclear reactors, and Nestlé, whose marketing of infant formula has been a major target for activists for years.

The chief executives recruited by Stephan Schmidheiny to the BCSD, to help him advise Maurice Strong, represented key industrial sectors. Their role is important because business was the only independent sector – unlike NGOs, women, youth, indigenous peoples, trade unions, and farmers – that helped pay the Summit's bills. Their lobbying had an important impact on Agenda 21. BCSD members claimed to be acting in 'personal, not institutional roles', but were so successful with their 'advice' that the only mention of corporations in Agenda 21 was to promote their role in sustainable development. No mention was made of corporations' role in the pollution of the planet, nor was there any kind of guidance or regulation to ensure that they are more responsible in the future. This success earned them the nickname of the 'Sustainable Council for Business Development'.

The BCSD worked closely with the ICC in promoting the idea to the Summit that economic growth, new technologies, and 'open and competitive markets both within and between nations' were key elements in solving environmental and developmental problems. In the fortnight's run-up to the Rio-Summit talks, Maurice Strong appeared at conferences in Rio de Janeiro and re-endorsed both organizations' principles for sustainable development. What is more, all other UN heads of agencies were persuaded to sign the principles with the single exception of Peter Hansen, head of the UNCTC, who was never approached with the document.[9]

Following the fourth PrepCom meeting in New York, these businesses were keen actively to present themselves as part of the solution to the global environmental crisis, rather than as part of the problem. Peter Bright, head of environmental issues for the UK oil company Shell, speaking on behalf of the ICC, told government representatives in New York on 1 April that Agenda 21 should take advantage of the enormous resources of businesses and go beyond the role of regulation.[10] And at a BCSD meeting on 29 May 1992, Strong told reporters that 'no assignment has meant more to us' than working with the

BCSD, as BCSD 'has become a cadre of the world's leading practitioners of sustainable development, [and] BCSD staff have become our happy partners with our own secretariat in this process'.[11]

MONEY MATTERS

Of course, Strong's words were not totally disinterested: Schmidheiny and other corporations had substantially helped finance the Summit and even the parallel non-governmental Global Forum. In fact, apart from lobbying heavily within the system, business actually helped pay for much of the Summit, a tactic from which they reaped considerable benefits. In particular, it helped defeat recommendations from within the UN – from the UNCTC as a matter of fact – that would have called for a much stricter monitoring and regulation of corporations, replacing them with the much weaker idea of 'self-policing'.

The Summit secretariat spent US$16.9 million in preparing for Rio. Like any other UN body, it would normally be expected to rely on UN funding plus any additional money it could raise from sympathetic governments, and most of its money did come from these sources. But almost a fifth of the financing for the Summit came from corporations.[12] The secretariat set up a Voluntary Fund to raise money directly from governments, which together with the funding from the UN's regular budget raised US$6 million. In addition, it set up a special US$10.86 million Trust Fund to which anybody could contribute and private corporations certainly did, through yet another fund called EcoFund. Without doubt, this was Strong's forte – raising private money and acquiring expertise to carry out government approved projects.

In 1990 when all the arrangements for the Summit were being made, Maurice Strong helped retired Washington, DC lawyer Benjamin Read set up EcoFund '92.[13] Registered as a non-profit organization, it had raised US$2.3 million before the New York PrepCom. Read told us that he had increased this to US$4 million by the time the Summit closed in Rio, and eventually anticipated topping it up to US$4.6 million when the final donations and royalties came in.

The biggest contributor to EcoFund was Swatch of Switzerland, owned by the Schmidheiny family. Swatch sold special Earth Summit watches and donated 5 Swiss francs for each watch, a total of US$1.8 million, i.e. over 10

per cent of the secretariat's costs of preparing for the Summit. Chemical giants like UK-based ICI, USA-based Minnesota Mining and Manufacturing (3M), and oil companies like USA-based Atlantic Ritchfield Oil (ARCO) were also recruited to pay into the Fund.

The details of the financing as told to us by Read were as follows. ICI and ARCO were quite small contributors, putting in US$25,000 and US$35,000 each. 3M put in US$100,000, while Coca-Cola put in US$200,000 plus two executives to help organize logistics in Rio. Coke also prepared 'Earth Summit Kits' for every elementary school in the USA and other English speaking countries. Other than Swatch, ICI, 3M, and ARCO, contributors to EcoFund included Fiat of Italy, Asahi Glass Co, Kadokowa Shoten Publishing, Kinki Nihon Tourist Co, and Ito Yokado Supermarkets from Japan. The EC also put some money into EcoFund.

EcoFund helped pay, among other things, salaries of Summit staff, Summit youth ambassadors, and an indigenous peoples meeting prior to the Earth Summit. Corporate contributors who paid over US$100,000 into EcoFund were allowed to use the 'In Our Hands' Earth Summit logo. At least four companies took advantage of the logo – Fiat, two companies that manufacture personal badges monitoring ultraviolet and acid rain levels, and Swatch. The secretariat later maintained that NGOs had the same right to use the logo for free, but no NGO was ever known to have used the logo. UN staff told us that this was not the first time that corporations had sponsored the UN – previous examples include Benetton's paying for guards' uniforms, and IBM's paying to revamp UNEP's computer system – but the scale is certainly unprecedented.[14]

What is more, some events were directly paid for by corporations. For example, Swatch sponsored a cultural gala and reception for negotiators at the New York PrepCom meeting. Meanwhile, other corporations gave in-kind support to the Summit. As mentioned earlier, Coca-Cola loaned the services of some of its top executives to assist the Summit in its promotional campaign and produced a series of public service ads promoting the Summit through its New York based advertising agency Lintas. Volkswagen gave a fleet of 'clean' cars for use by the Summit secretariat and delegates. Xerox donated equipment to the conference organizers.

Furthermore, Strong got corporations to pay for a private newspaper that he helped set up at the Summit – the *Earth Summit Times*, now called *Earth Times*. This made its appearance at the New York PrepCom but at that time

advertised itself as an independent newspaper. This changed at the Summit, when the masthead clearly declared it to be the official newspaper of record. According to Paul Hoeffel, editor of the UN's own development newspaper – the *Development Forum* – Maurice Strong sidestepped the UN's own Department of Public Information to ask journalist and entrepreneur Pranay Gupte to set up the *Earth Summit Times*. Gupte set up the paper with money from businesses.

Equally insidious was the fact that the BCSD's public relations consultants, Burson-Marstellar, offered to help the Summit with its public relations. According to Jean-Claude Faby, director of the Summit's New York offices, it offered to do all the Summit's public relations, but when the secretariat said it was too expensive, the head of its US operation offered to do the work *pro bono publico*. As far as we can tell, the secretariat decided not to accept this offer. To recall, Burson-Marstellar has extensive experience in presenting environmentally friendly images for corporations. It helped Exxon present the best possible face after the *Exxon Valdez* disaster in Alaska and did the same for Union Carbide after Bhopal. It also helped to stem the negative publicity surrounding the problems after the Three Mile Island nuclear reactor failure. Attempts to discover the full extent of corporate sponsorship in UNCED and how it was conducted were frustrated, however, by the lack of information.

Again, in our view, business and industry are not to be blamed for having sponsored UNCED and taking advantage of it. They were basically profiting from an opportunity offered on a golden plate. However, they must be criticized for double-speak, and for using the Earth Summit as a strategic event without being willing even to consider the profound changes that would be necessary in order to take significant steps towards a sustainable society. Indeed, many of the corporations that paid for the Earth Summit had appalling environmental management records. Perhaps more insidious still, many of these corporations funded anti-environmental lobbying groups in the United States and probably elsewhere. In short, while promoting themselves through the Earth Summit as the solution to the environmental and developmental problems, they simultaneously opposed environmental protection standards and legislation at the national and the local levels. This is what made business and industry lose credibility and legitimation as serious agents in dealing with the global environmental crisis. This is what turned their sponsorship of UNCED into a greenwashing farce.

In our view, it is mainly Maurice Strong and the UNCED secretariat that have to be blamed for giving TNCs a comparative advantage over other independent sectors and for deliberately presenting them as the solution to the crisis. Governments must be blamed for having abdicated their responsibility and often for actively supporting this perversion of the UNCED process. Most NGOs, finally, deserve blame for not having been perceptive enough to notice what was going on, notably at their expense. This is even more embarrassing as funding of the so-called NGO-event, the Global Forum, is even more mysterious and impenetratable than is the funding of the Earth Summit.

8

CHANGING WHAT?

As we have seen, the UNCED process in general and the Rio Conference in particular have led to the promotion of business and industry and their worldview over other agents and their worldviews. We have also seen that there are several reasons for this. These are:

- The set-up of the UNCED process as a lobbying exercise, where everyone defended his or her interests, as opposed to a collective learning endeavour;
- The fact that business and industry understood Rio precisely as that: a lobbying and public relations effort, thus using UNCED to present themselves as the solution rather than as the problem;
- The fact that Maurice Strong and the UNCED secretariat were actively promoting business and industry and their views over other agents. As a matter of fact, neither Strong's nor the Brundtland report's views differ much from the view of business and industry;
- The fact that governments have basically abdicated their responsibility: many Northern governments have become the spokespersons of Northern business, while Southern governments were advocating more development and economic growth, thus playing into the hands of business;
- The fact that the environmental movement – the only potential counterforce – was highly fragmented, organizationally and ideologically, and that the UNCED set-up enhanced, not reduced, this fragmentation. This includes the fact that nobody in the movement seemed to be able to see that UNCED weakened not strengthened their movement; and finally
- The absence of any intellectual leadership in and around UNCED. There

was not and still is no alternative vision sufficiently coherent and strong
to face the 'Brundtland' ideology presented in Chapter 1.

All these reasons led to the result that the business and industry worldview
came out of Rio as the solution to the global environmental crisis and no longer
as its cause. Schmidheiny and his BCSD book entitled *Changing Course* became
as important an output of the process as the Brundtland report was an input
to it.[1] And their views of the problem are hardly any different.

This chapter is divided into three sections. In the first section we present
the core views of the BCSD and the suggestions it makes for solving the
environment and development problems, as outlined in the book published
right before the Summit. In the second section we present and discuss the
discourse that accompanies these core views. We do not address here the
suggestions made by the ICC, which published a summary of its ideas in book
form as well. This was titled *From Ideas to Action* and was co-authored by Jan-
Olaf Willums and Ulrich Gölüke.[2] In the introduction to this book, the authors
stress that the difference between *Changing Course* and *From Ideas to Action* is that
the former is a book about vision and a call to action. The latter was intended
to supplement the former as a handbook on how to transform the vision into
reality. In the third section we examine this view and outline why we think it
is flawed and not suited to deal with the global environment and development
crisis.

The two initial sections of this chapter correspond to the distinction in
Schmidheiny's book between two different themes: there is first a core in
which environment and development are considered from the perspective of
organizational development and change. This core actually constitutes the bulk
of the book (Chapters 6 to 17), and we basically agree with it. But there is a
second theme, which is the discourse on environment and development. This
discourse, i.e. the BCSD's ideas on sustainable development, is in our view
basically flawed.

THE CORE: MANAGING
ORGANIZATIONAL CHANGE

In order to understand what the core of Stephan Schmidheiny's book is all about, we have to place ourselves within the context of recent management philosophies, in particular the philosophy called total quality management (TQM).[3] The two main features of TQM are its customer focus and the fact that a product is looked at in its overall production process from its inception until it ends up with the customer. Both, process and customer focus combined, lead TQM to imply changes in corporate strategy, output, work, people and their training, as well as leadership, organizational architecture, and organizational culture. Most of the big corporations such as the ones that are represented in the BCSD have undergone quite substantial internal reorganization efforts over the past few years, inspired by TQM and similar management philosophies.

To be sure, TQM has nothing to do with environmental problems. It is an integrated means of dealing with a changing corporate environment, perceived essentially as globalization, acceleration of technological change, cultural fragmentation, and individualization of the customer. To all these changes corporate organizations respond with a focus on the management of their overall process, by trying to integrate into one single approach the hard and the soft factors of a company, i.e. technology and organizational structure on the one hand, and organizational culture and learning on the other. Ultimately, TQM is to make the organization or the company more efficient, i.e. more fit for a changing environment. But such fitness is not going to be achieved by technological change alone. It is equally the result of changes in the organizational architecture, i.e. the organization's structure, its leadership, and its culture. In short, TQM asks the company to focus on the customer – who generally only sees the output -- and rethink its production and management process accordingly.

Interestingly, the environmental challenge is not alien to but actually reinforces the TQM philosophy, focused as TQM is on the results (outcomes) and the efficiency of the process. Indeed, the natural environment is interpreted within the TQM framework as analogous to the customer: what the low quality product is to the customer, pollution is to the environment, i.e. basically a sign of organizational inefficiency. Says Stephan Schmidheiny:

'More and more companies are realizing that the pollution they produce is a sign of inefficiency and that waste reflects raw materials not sold in the process'.[4]

The environment, like customer satisfaction, is basically considered to be located on the output side of the production process, yet starting from this output side environmental and customer considerations must be inbuilt, according to TQM philosophy, into the whole process and more generally into the whole organization, as well as into its strategy. 'The environmental considerations must be fully integrated into the heart of the production process, affecting the choice of raw materials, operating procedures, technology, and human resources. Pollution prevention means that environmental efficiency becomes, like profitability, a cross-functional issue that everyone is involved in promoting'.[5]

Schmidheiny then gives a series of examples of how the environmental focus leads to changes in the organization and the production process in various companies. Unfortunately, these examples remain unanalysed and sketchy. Others have gone much further in conceptualizing organizational change as a result of the new environmental focus. Dyllick, for example, distinguishes six foci of change, namely product development, materials acquisition, production, marketing/sales, logistics, and recycling, each of which must be addressed at the four following levels, i.e. infrastructure, personnel and organization, management, and communications/public relations.[6]

In short, TQM and other related new management philosophies have made companies focus on some of the environmental consequences of their production process by analogy with their rediscovered focus on the customer. As a result, environmental problems are framed in terms of organizational efficiency. We indeed detect a lot of honest and laudable management efforts to deal with the environment in this way and are all in support of them. However, TQM and the analogy between environmental and customer focus have some limits and flaws.

First, TQM, and even more so Schmidheiny, sees the production process as separate from the customer and by analogy from the environment. What the customer and the environment 'see' is the output, i.e. pollution and waste in the case of the environment. Consequently, the production process and the organization have to be designed so that they are more efficient, i.e. in a way that reduces such undesirable output, ranging from waste reduction, to

recycling, to pollution prevention. Thus, it is not the idea of industrial production that is questioned, but some particularly nasty outputs. The philosophy, therefore, does not go to the roots of the problem.

Second, the unit that has to change remains the organization and the company, not society. The main concern is profit maximization and in some cases survival. But it is always survival of the unit, not of society or the environment. The problem is that an organization or corporation is not a self-sustaining unit. Current management philosophies, even if they are focused on the environment and the customer, cannot by definition go beyond this idea. What is beyond it generally turns into a lofty discourse on (business) ethics.

Third, there is a flaw in the analogy between the customer and the environment. Even in the case of the customer, TQM in particular and the economists in general are probably wrong, as customers are not organized enough to speak up for themselves, fragmented and manipulated as they are. But the problem is even more serious in the case of the environment: who speaks up for the environment when a company does not treat it satisfactorily? In short, how detrimental the outcomes of a given production process are for the environment remains defined by the company, not by the environment.

There are three types of agents who, according to the BCSD, can speak up for the environment, namely:[7]

1. Governments via regulations (command and control);
2. Governments via their intervention in the marketplace (economic instruments);
3. Business itself (self-regulation).

And we should add here a fourth:

4. The market through consumers' preferences.

Schmidheiny is clearly not in favour of command and control government regulation, such as emissions and immissions standards, though he thinks that in case of urgency such regulations might be justified. He is somewhat lukewarm about economic instruments such as pollution taxes and charges, tradable pollution permits, resource quotas, etc. From an organizational perspective, economic instruments are quite logically preferable to command and control, since they allow for more flexibility and innovation in the organizational change process. Not surprisingly, Schmidheiny is most in favour

of self-regulation, which he immediately ties to the leadership business will take in solving environmental problems. He says: 'It is time for business to take the lead, because the control of change by business is less painful, more efficient, and cheaper for consumers, for governments, and for businesses themselves. By living up to its capabilities to the full, business will be able to shape a reasonable and appropriate path toward sustainable development'.[8]

However, this opposition Schmidheiny points out between government and business, i.e. between command and control regulation by the state and self-regulation by business, is in our view a false dichotomy which obscures the real issue. The fact is that governments and businesses are not enemies, but allies, interested as they both are in economic growth and industrial development. There is indeed a collusion of interests: governments will only go as far in regulating business as regulation will not cut into GNP, which is ultimately where governments derive their income. Also, governments, like self-regulating business, have an interest in stimulating a type of technological progress that will give national industries a comparative advantage. The only difference between government and business might well be that government is slower than business. The picture that Schmidheiny tries to give of business being the victim of government regulations is therefore misleading.

And this goes hand in hand with yet another picture that permeates the entire book, namely that business is the humble servant of the consumers, a picture that is, by the way, implicitly promoted by TQM philosophy. Although this picture might be true in the case of small and medium sized businesses, we have some doubts that this is the case for multinationals, such as the BCSD members. Many of them are in quasi-monopolistic positions and self-regulation, in this case, will be more or less equal to self-interest.

In the absence of competition, of government regulations and of consumers' and citizens' pressure, the BCSD and the ICC have nothing else to propose than 'business ethics', that is, highly abstract and in any case non-binding principles such as the ones compiled in the ICC Business Charter for Sustainable Development. Of course, we are not opposed to principle 10, for example, which is all in favour of the 'precautionary approach', nor do we have a problem with principle 15, 'openness to concerns'. Other principles, though, are more questionable, such as the promotion of technology transfer. Overall, we do not believe that such lofty principles are up to the challenges of preventing the type of global environmental degradation many TNCs are

causing. More generally, we do not believe that ethics can slow down, let alone redirect, organizational behaviour where stakes and interests are as high as they are in TNCs, in governments, in science, or in the military, for example. Quite interestingly, almost all examples of self-regulation given by Schmidheiny, as well as all the forces that drive industry, according to him, to self-regulation, stem from some sort of outside pressure on business and industry.

Moreover, this collusion between governments and businesses is even higher at the global level than it is at the national levels, given the fact that at the national level, at least in the North, there is some tradition of separation of power. By pushing global concerns and global approaches to these concerns as actively as business and industry do, they simultaneously push towards the creation of a new global reality on which TNCs have a better handle than on the national and the local realities. Without institutional precedent, the global reality is more favourable to business, at least to big business, than other levels of society. And this is implicitly what the move towards sustainable development means, according to the authors of *Changing Course*: being in favour of sustainable development in and around UNCED is probably just another means of accelerating the erosion of the national dimension and promoting the global approach where TNCs do have a comparative advantage over governments.

THE DISCOURSE: ECONOMIC GROWTH AND FREE TRADE

We have seen that the core views of the BCSD are quite in line with the newest management thinking. Such thinking leads business and industry to deal with environmental problems as an issue of total quality, i.e. ultimately as an issue of organizational efficiency. However, the evolution of such thinking is quite unrelated to UNCED. As such, it is also quite unrelated to the global environmental crisis, and the corresponding challenges. As a matter of fact, the BCSD has actually taken advantage of the forum UNCED provided to display its environmental and other management efforts. As we have seen in the previous chapters, it has also used the UNCED forum to promote its views as a solution to the global environmental crisis. Our criticism is that the inference Schimdheiny and others make is wrong: even if

environmental and other management efforts may be quite appropriate answers to changes in the global marketplace, one cannot deduce that the same environmental and other management efforts are also a solution to the global environmental crisis. Quite the contrary is the case, as we will see: environmental management efforts that provide a given company with a competitive advantage in the global marketplace might well be counter-productive for society or the planet overall.

As in the case of the Brundtland report, Schmidheiny's analysis of the global environment and development crisis is fundamentally flawed. First, the cause of today's crisis is not seen as being industrial development, but is attributed to humanity and humans, i.e. the very 'nature of human activity'. Moreover, Schmidheiny does not seem to have a sense of the globalness of the problems, and therefore not of their urgency either. Economic growth is an imperative, a 'requirement' as Schmidheiny says.[9] And 'economic growth does not necessarily hurt the environment'.[10] It is synonymous with 'human progress';[11] in any case it is the answer to growing needs. There are virtually no input limits to economic growth,[12] whereas output problems are not seen as being global in nature. In line with TQM philosophy, they are problems of pollution and waste, i.e. precisely the type of problems industry can deal with. In any case, everything is a matter of efficiency and efficiency is what environmental management is all about: the efficient use of resources and the decrease of pollution, both being challenges for management and technology development. The best way to achieve such efficiency is not government regulations, but open markets, albeit markets that reflect the cost of environmental degradation, i.e. markets that 'internalize externalities'.[13] Quite logically, therefore, the solution to all our problems is not to be found in UNCED, but in GATT. Says Schmidheiny: 'Perhaps the most effective way forward is to improve the ability of GATT to minimize trade interferences caused by environmental regulations'.[14]

New open and competitive markets certainly profit TNCs, especially monopolistic ones. But do open and competitive markets solve the global crisis or, to begin with, reduce use of resources and environmental pollution? In a review called *Changing Course*, the *New Scientist* quotes World Bank economist Herman Daly as saying that in fact competition through a combination of open markets and multinational corporations results in a 'permanent international standard-lowering competition to attract capital. Wages can be lowered as can

be safety standards and environmental standards – all in the name of reducing costs'.[15]

Now, corporations will say that they manufacture in Southern countries because the cost of labour is more competitive and regulations are laxer, which theoretically will profit the consumer. But as Daly says, read lower for competitive. Will the consumer object? From what we know about consumer behaviour, the answer is no. In other words, if the regulations are too strict and the wage costs too expensive, the companies will leave. The consumers will continue to buy, especially in the case of monopolistic TNCs. Therefore, open and competitive markets will not only lead to the lowest common denominator, read lowest environmental standard, but moreover such low environmental standards will retard, if not prevent, the internalization of externalities, i.e. the hope that the market will ever get the price right.

In short, Schmidheiny's book contains two themes, the promotion of environmental management in line with TQM philosophy on the one hand, and a discourse on economic growth, free trade, and open and competitive markets on the other. They are unrelated. Moreover, neither this discourse nor environmental management will help solve the global environmental crisis: if economic growth and open markets lead to more and not less environmental degradation, environmental management will at best slow down resources consumption and pollution. But we suspect that the hidden agenda behind promoting environmental management is not to stem global pollution, but rather to find another competitive advantage in the global marketplace. Says Schmidheiny: 'Yet ultimately we have to accept that a move towards sustainable development will cause far-reaching change in the structures of business and industry; there will be losers and there will be winners'.[16] In this way, UNCED was used by the BCSD, ICC, and many others to present their latest management efforts to stay competitive as being the solution to the global environment and development crisis. By connecting environmental management to economic growth and global trade, they could, moreover, rehabilitate business as a means of environmental protection.

'CHANGING LABELS' — A CRITIQUE

It comes as no surprise that, like the EcoFund corporations, the BCSD members are not all environmentally friendly themselves. Among them we have Norsk Hydro AS, Asea Brown Boveri, 3M, Ciba-Geigy AG, Aracruz Celulose SA, Mitsubishi Corp, Shell, du Pont, Tata Industries Ltd, Browning-Ferries Industries, Dow Chemical Company, and others. And that might be one of the reasons why they joined the BCSD to begin with. Mitsubishi, for example, is one of the world's leading destroyers of tropical rainforest,[17] while du Pont Corporation is the inventor and largest producer of ozone-destroying substances,[18] and Brazil's Aracruz Celulose is the world's largest exporter of bleached eucalyptus pulp.[19]

Indeed, we see a certain contradiction between the new environmental vision of the companies which are profiled in the second part of Schmidheiny's book as environmental leaders on the one hand and their actual doings on the other. We will rely here on Greenpeace material which has already highlighted some of these contradictions. Indeed, the BCSD was thwarted somewhat by Greenpeace, which scored two major coups. First, it pre-empted the launching of *Changing Course* in May 1992 with its own critique a few hours earlier. Then, in Rio, it launched its own book called the *Greenpeace Book of Greenwash*,[20] a detailed critique of nine of the BCSD corporations, the day before the BCSD had its gala pre-Summit meeting. Finally, on the day of the meeting, several Greenpeace members slipped into the press briefing where Maurice Strong was enjoying a photo-opportunity with Schmidheiny and other BCSD members like Erling Lorentzen, the Norwegian chairman of Aracruz, and Frank Popoff of Dow Chemicals. To their consternation and the amusement of the press, a Greenpeace television reporter announced that, as they spoke, Greenpeace's flagship vessel, the *Rainbow Warrior*, was blocking Victoria, Aracruz's main export harbour 190 km north of Rio to protest against the BCSD and its takeover of the Summit.[21]

The Greenwash report, prepared by Kenny Bruno, attacks Aracruz for the very example cited in *Changing Course*, i.e. reforesting the Amazon which the book claims taught the company that 'enlightened environmental and social stewardship can be combined with corporate profitability'.[22] The company claims it took over a devastated, unproductive, deforested area and reforested it with highly productive fast-growing eucalyptus that could be harvested every

seven years. Bruno, however, contends that Aracruz's own environmental impact statements show that 30 per cent of the region had regenerated second-growth forests that were cut and replaced with monoculture eucalyptus forests.

Aracruz claims that eucalyptus forests are part of sustainable forest management, but Greenpeace points out that on the contrary eucalyptus monocultures destroy topsoil, water tables, and biodiversity. Aracruz does not mention either that the land it now harvests belonged to the Tupiniquim native peoples and was handed over to Aracruz Celulose by the former military regime. Aracruz used their food production lands to grow timber on. At the press conference launching the 'Greenwash' book, Greenpeace Brazil representative José Augusto Padua said that Aracruz was fined twice the previous year for not complying with environmental laws.

Then *Changing Course* highlights a Mitsubishi project on sustainable forestry in Malaysia on a laughably small 50-hectare university plot. As Fred Pearce points out in the *New Scientist*, logging companies in Malaysia annually fell 450,000 hectares of primary forest every year.[23] Another more bizarre attempt at greenwashing includes a comic book distributed to all Japanese high school students that depicts Hino, a fictional Mitsubishi executive, who travels around the world to find out the truth behind the corporation's bad public image and discovers that, far from being the major cause of deforestation, poor peasants are the real problem.[24]

One more example will suffice. In *Changing Course* du Pont cites its chief executive officer, Edgar S. Woolard Jr, who took over the company in 1989, as an example of how 'committed leadership from a chief executive can unleash a cascade of environmental improvements throughout the corporation'.[25] In fact, Bruno points out that the company invented and manufactures the largest quantity of CFCs in the world, and is currently believed to be the leading ozone depleter. And in 1989, after the Helsinki declaration ordered the phase-out of the chemical by 1995, du Pont lobbied against a faster phase-out, while two years later its management blocked a resolution from some of its own shareholders to phase out the chemical by 1995. In contrast, *Changing Course* claims that its leadership has been 'precautionary' and 'proactive', because it called for a phase-out of CFCs in 1988 and gave a deadline of the year 2000.

Given what many of the companies highlighted in Schmidheiny's book really do, the book appears as a public relations exercise, a form of greenwash, or

as it was said in Rio 'changing labels' as opposed to 'changing course'. Now, such inconsistencies between saying and doing are not in the interests of the companies either, as they throw doubts on their credibility. Of course, the companies will say that they are in the process of changing, that this is only the beginning, and that they will implement environmental management. Besides the fact that, as we have seen, environmental management is not the solution to the global environmental crisis, every single step towards environmental management of TNCs has generally occurred, at least as far as now, in response to outside pressure.

A UNCTC survey conducted specially for the Summit discovered that the BCSD recommendations were quite the opposite of why companies were changing in the first place.[26] Legislation and not self-regulation is generally the driving force behind a change in corporate environmental policy. The UNCTC Corporate Environmental Benchmark Survey says: 'Changes in home countries legislation were cited as the most significant factor in influencing the companies' environmental policies and programmes on a country-wide basis'. Of all companies surveyed by the UNCTC, 59 per cent noted that a change in home country policies provoked a company-wide policy change. A survey conducted by Tufts University in Massachusetts also identified government laws and regulations as the most influential factors in corporate environmental policies.

But the UNCTC and Tufts were not the only ones who found that government agents direct corporate environmental policy. A curious fact was brought to light by Friends of the Earth, which obtained a draft copy of *Changing Course* and discovered that the BCSD's own research had come up with the same conclusions. Apparently, a survey commissioned from the accounting firm of Deloitte and Touche by the BCSD reported that 'government regulation, either already existing or thought to be on the horizon, was often cited as the most powerful force, encouraging the generation of this type [i.e. environmental] of information'.[27]

On the other hand, international guidelines for corporate behaviour were not very widely used. The UNCTC survey showed that 'over half of the respondents were found not to utilize international guidelines. Many transnationals were unaware of the nature or existence of particular international guidelines'.[28] Of the twelve international guidelines listed, the least frequently followed were UN guidelines (i.e. UNEP, FAO), followed by ICC guidelines

— except in Europe where ICC guidelines were known and followed by quite a number of companies. This and the preceding data indicate that strong national legislation is perhaps the most effective way to shape corporate policy.

There is, indeed, enough evidence to support the contention that almost all changes towards environmental management were initiated by some sort of governmental regulation. For example, industry said that 'market based incentives' were a substitute for regulation that would help keep the environment clean. Many of the big NGOs have fallen for this line. Thus, GM says that a system of 'pollution credits' is a market incentive, because it gives corporations the right to trade pollution 'rights'. If we think of this as a market-led incentive, then we believe that we simply need to allow the market to correct itself or perhaps help correct itself. Yet the reason that companies start trading pollution 'credits' or 'rights' is because of government regulation, in this case the US Clean Air Act of 1990 which they fought tooth and nail but which now forces them to reduce or face fines. We should not forget that the real force of change in the past has been legislation, i.e. government interventions and, most importantly, public accountability. And this is most likely going to be true also of 'full-cost pricing' and 'environmental reporting', the two suggestions of self-regulation made by Schmidheiny.

The problem, however, is not necessarily TNCs. The problem lies in the fact that any organization will want to control the very forces that try to limit and shape its activities. This is especially true of big multinational corporations, which actually do have the power to influence the very processes by which they are regulated. The UNCED is a perfect illustration of this: it had the potential to regulate global business and industry, especially TNCs, since industrial development, at least in the beginning, was seen as causing global environmental problems. But the BCSD, with the aid of the UNCED secretariat, made sure that this would not happen.

Governments were, indeed, offered alternative ideas on business, industry, environment and development, drawn up by the UN itself. These alternatives were elaborated at the suggestion of another UN body, ECOSOC, which commissioned the UNCTC to draft a set of recommendations for the regulation of multinationals. The G-77 block of Southern countries asked that these ideas also be taken to the New York PrepCom. Harris Gleckman of the UNCTC helped draft and redraft ideas to make them acceptable to the secretariat, but was reduced to lobbying individual governments at the New

York fourth PrepCom, when the secretariat displayed little interest. After intensive lobbying, some governments like the Swedish and the group of 77 Southern countries did agree to take up the UNCTC proposals, but they were defeated by the big powers, the USA and Britain in particular, during the government negotiations. At the same time, according to Greenpeace, the ICC was actively lobbying the Swedish government in Stockholm to withdraw the call for TNCs to internalize environmental costs in their accounting and reporting processes. At this point, UNCTC was fighting a last-ditch battle because only weeks before the New York PrepCom Dr Boutros-Ghali, the new UN Secretary-General, had cut the ground from under its feet by axing the department.

The UNCTC suggestions were completely ignored both by Maurice Strong's secretariat and Stephan Schmidheiny. In fact, Schmidheiny told the authors at a press conference in Rio that the UNCTC proposals for regulation were no concern of his.

CONCLUSION

The UNCED process in general and the Rio Conference in particular were a unique platform for the Business Council and the International Chamber of Commerce to present their view. It is a view of a particular management philosophy, accompanied by an overall apology for the free market, a view that proposes self-regulation at a time when governments' legislative authority and legitimation are being eroded. We have seen how this became the dominant view at Rio, and will probably remain so for a certain time to come. But to conclude, we offer a critique of this view, as we believe it will aggravate, not alleviate, the global environmental and developmental crisis.

Given that this view is rooted in TQM management philosophy, the environmental challenge from the perspective of a company is basically a problem of efficiency. The company must be made eco-efficient. What contributes to this is more efficient management, more efficient organization, a cultural change in the organization, and of course technological improvements. The environmental problem, at the company level, is thus redefined as a technical problem. This is furthered by the focus on more efficient resource use and better pollution control. There is an underlying assumption, in

Schmidheiny's book and elsewhere, that the most efficient company does not pollute. By extrapolation, Schmidheiny seems to believe that economic growth can be 'decoupled' from environmental impact.[29] We oppose this view of techno-efficiency for three reasons.

First, technology and efficiency are a means, not an end. Elevating efficiency and technological solutions to become the goal itself, as Schmidheiny does, will fail, because it promotes a technocratic management approach to a problem – the global crisis – which is not fundamentally technical in nature. We therefore differ from Schmidheiny in the very assessment of the crisis: in our view, this crisis is the result of the industrial civilization, whose origin can be traced back to the Scientific Revolution of the sixteenth and seventeenth centuries in Europe. Trying to solve the crisis with the very tools that created its origins will only accelerate the problem. Rather than ever-better technology, ever-better management on an ever-more-global scale and at an ever-faster pace, in our view the way out is to slow and to scale down. The direction to go is deindustrialization, while building up local and regional communities to manage their own socio-economic activities and resources. Moving in this direction implies a collective learning process which is not primarily a matter of efficiency.

Second, we oppose this technological eco-efficiency solution on cultural grounds. If this eco-efficiency approach makes perfect sense at a company level, where it will certainly lead to increased competitiveness, it will, however, at best slow down the acceleration of global environmental degradation on a planetary scale, but not reverse it. Moreover, on a societal level it is counterproductive. The reason is that eco-efficiency is a Northern approach. It is an approach particularly geared to reducing pollution problems for which there are indeed technological solutions. But in the South environment and development problems are of a different nature. There are, of course, pollution problems in the South as well, but these have been imported by the North and Northern industries. The primary problems in the South are resources problems. By this we mean problems of access, control, participation, and governance of natural resources at local and regional levels. And this is primarily an equity and not a technological issue. By declaring the eco-efficiency approach to be universal and exporting it to the South, Schmidheiny, the BCSD, the ICC, and UNCED are guilty of ethnocentrism, in addition to promoting inadequate solutions.

Third, we oppose this eco-efficiency approach because by promoting eco-efficiency planetwide, accompanied by the call for open competitive markets and full-cost pricing, an economic rationality is being extended to everything that previously had social, cultural, and natural values attached. Everything – nature, culture, beliefs, and values – will have a price tag and will be judged on whether or not they contribute to eco-efficiency. The planet-wide extension of economic rationality under the cover of eco-efficiency will therefore further cultural destruction and erosion. It will promote the ideology of rational choice with the self-interested individual at its core, and destroy the remaining cultural restraints on individualism. It will destroy the local by imposing upon it a global market rationality. It will eventually wipe out the very cultural forces from which ways out of the present crisis could eventually emerge. In short, the price of global eco-efficiency, imposed by the planet-wide extension of economic rationality through open and competitive markets, will be further cultural erosion. The cultural consequences of this evolution are difficult to assess but certainly go in the direction of growing individualism and fundamentalism. Even if eco-efficiency were a solution to our environment and development problems, it would probably have to be rejected because its cultural consequences are so disastrous that it is not worth the price.

UNCED and its reference to a planetary urgency was invoked by Schmidheiny, the BCSD, Strong, many governments, especially in the North, and even NGOs such as IUCN and WRI to promote global eco-efficiency, further economic growth, universal economic rationality and open competitive markets as the answer to the environment and development problems as they see them. After UNCED, this discourse and view are now dominant. The BCSD has significantly contributed to legitimizing this view, thus promoting a new global reality where environment and development problems are supposed to be solved. Establishing this global reality as a legitimate one has allowed business and industry, especially TNCs, to free themselves from government control – under which they certainly were at the national levels – and become legitimate global agents. At best, the governments are now partners of business and industry in this overall global environmental management scheme, a fact that was further cemented in the Rio financial and institutional arrangements.

Part IV

FINANCE AND INSTITUTIONS

This final part will deal with the UNCED outcomes and follow-ups, which were separated at UNCED into a financial and an institutional aspect. It was at the fourth PrepCom in New York that the issue of the follow-up to the much anticipated Rio agreements on climate change, biodiversity and Agenda 21 came to the fore. Days before the PrepCom started, Maurice Strong kicked off the discussions by telling reporters that a lot of new money would be needed to finance the draft plans for saving the planet. He said that the secretariat estimated that it would cost US$125 billion a year in new aid between 1992 and the year 2000 for Southern countries to clean up their environment, US$70 billion more than the current global total of bilateral and multilateral 'aid' that was being sent to the South. When a Reuters correspondent asked how much the total cost would be – aid plus local financing – Maurice Strong said perhaps US$600 billion a year. There were of course some gasps. Others in New York at the time, like the British minister for trade and industry, Michael Heseltine, questioned the likelihood of anything near that sum being raised, although cynics pointed to the fact that as much was spent on the previous year's Gulf War. Parallel to these 'finance' discussions, governments also began serious talk of a monitoring mechanism for the Rio agreements, which they called the 'institutions' discussions.

As we have pointed out all along in this book, the global environmental crisis was being reframed through UNCED as a development problem, development essentially being needed in the South. Be it in the Brundtland report or in Schmidheiny's book, the solution to the global environmental problems is said to be efficiency, i.e. technology and investment in technological progress as well as

in economic growth in the South. In other words, solving the global environmental crisis basically boils down to a matter of money. By the fourth PrepCom in New York, governments had come to agree on this. Jamsheed Marker, head of Pakistan's delegation to the summit and the chairman of G-77, said: 'What we want is a credible commitment on financial resources so that we do not leave Rio with a mere statement of good intentions and wait to see how it is going to be implemented.' And his partners in the wealthy North agreed. Curtis Bohlen, head of the US delegation, said in New York: 'The US accepts that if the world is to fully achieve sustainable development, industrialized countries must generate new and additional financial resources.'

Global environmental problems now turned into the question of how this money was going to be found and allocated. This question became the source of much contention between the countries of the North, which were viewed as the source of money and wanted to retain control of it through institutions that they controlled like the World Bank, and countries of the South, which were viewed as the recipients of the money and wanted an equal say in its distribution, perhaps through new institutions. Two weeks after the end of the New York talks, a meeting of 'eminent persons' was called in Tokyo by the former Japanese prime minister, Noburo Takeshita. Officially it was a private meeting, but unofficially it was set up by Maurice Strong's conference secretariat and according to senior members of the secretariat everybody was there with the approval of their home governments. The purpose was to iron out the massive disagreements on finance and get some commitments. Among those involved in the last minute negotiations were former US President Jimmy Carter, and former World Bank presidents Barber Conable and Robert McNamara. Also included were the finance ministers from Brazil and Pakistan, Marcilio Moreira and Sartaj Aziz (the conference host and the spokesperson for Southern countries respectively), as well as the heads of the African and Asian Development Banks. At the meeting itself the 'eminent persons' said that they thought US$10 billion was essential to get the ball rolling and that this could be financed through new economic instruments.

However, a month and a half later in Rio there were few commitments on additional money. The Rio speeches of some were spiced with slightly vague but tantalising offers of new aid as a precondition for 'sustainable development'. British Prime Minister John Major told his peers in Rio that 'Britain will mobilize its aid programme to back Agenda 21' and that 'money is the root

of all progress'. Chancellor Helmut Kohl's plenary speech called on the industrialized nations to help developing nations and said he aimed to increase German aid to reach the UN target of 0.7 per cent of their GNP. US President George Bush declared in his speech in Rio that developing countries 'will need assistance in pursuing . . . cleaner growths' and announced that Washington would increase its funding for international environmental initiatives by two-thirds. In terms of actual money, however, independent estimates figured that heads of state had promised US$2 billion extra in aid for the South. This was peanuts compared to the US$70 billion that Maurice Strong's secretariat had estimated needed to be spent. Although governments were not ready to commit themselves to anything near this sum, they did indicate how they would channel new money. They agreed in principle that the money should be delivered through the new Global Environmental Facility (GEF), the International Development Agency (IDA) at the World Bank, and through bilateral aid between individual Northern and Southern governments. Rich countries were supposed to try to bring this level of aid up to 0.7 per cent of their GNP. We deal with the financial aspect of the Rio follow-up in the next chapter.

At Rio the 'means' – i.e. the money – was separated from the 'substance' – i.e. the institutional follow-up. Governments agreed to set up an UN Commission on Sustainable Development (UNCSD or CSD) to monitor their decisions and make recommendations on financing the agreements. But they were unable to agree on the precise details of how the Commission would work and eventually decided to leave that for the UN General Assembly later in the year. These institutional aspects of the UNCED follow-up are dealt with in chapter 10.

CAN MONEY SAVE THE WORLD?

It was not mainstream thinking only that focused on money during Rio. Money was also what the Southern élite wanted. Indeed, as McCoy and McCully say: 'Although they initially saw the "environment" element of UNCED as a threat to their "right to develop", Southern governments soon realized how the North's stated environmental concerns might be used to extract economic concessions'.[1] Lobbying from Third World NGOs, especially the TWN, played a major role in persuading their governments of the advantages of this strategy, which was supported by most Northern NGOs. The Third World NGOs also played an important role in getting Southern governments to adopt a common negotiating position under the umbrella of G-77. China, which is not a member of G-77, normally supported the G-77 position. The G-77's arguments were summed up by its chairman, Ambassador Jamsheed K.A. Marker of Pakistan, who told the fourth and final PrepCom that 'the major cause of the continuing deterioration of the global environment is the unsustainable pattern of production and consumption, particularly in the industrialized countries', and that therefore 'developed countries must provide the major part of the resources required for sustainable development'.[2] The 'sustainable development' of the Third World should be paid for by the unsustainable development of the First.

The Northern governments, of course, did not publicly refute this call for money and financial aid, because it suited their thinking as well as their agenda. In line with Brundtland's and Schmidheiny's view that economic growth and eco-efficiency were going to solve the problems, the North's main strategy during the UNCED process was never to raise the question of the unsustainability of industrial development. Keeping the media and NGOs

focused on Southern demands for aid and technology transfer helped convey the message that environmental problems occurred mainly in the Third World and were caused by poverty, overpopulation, a lack of Western capital, management and technology, and insufficient application of currently fashionable Western economic theory. What is more, combining the threats posed by the poor Third World masses and the global environmental problems meant that a security syndrome was being created: the global environmental crisis became, at least for the North, a security issue, and when security is involved, the traditional problem-solving mechanisms are never questioned. Although they got little new money, Southern élites, like those in the North, benefited from the 'aid for sustainable development' and 'environmental security' arguments. Both helped distract from those showing that the only answer to the global crisis lay in profound structural changes, accompanied by deindustrialization and demilitarization. Groups and movements demanding agrarian reform, local control over resources, an end to large scale development projects, greater participation in decision-making, and a restructuring of global trade and finance were sidelined by the focus on the financial needs of Third World governments.

THE RIO CHEQUEBOOK

At or just after the Rio Summit, the World Bank and UNDP were the major sources of the new 'sustainable development' finance available. But as we have mentioned earlier, there was a lot of talk of new forms of bilateral aid at the Summit itself and if that funding does take off it could be equivalent to or even larger than the multilateral funding.

The USA was probably the first to come up with concrete commitments of bilateral money when, at the beginning of the New York talks in March 1992, it promised US$150 million over two years for Southern countries to study their emissions of global warming. In Rio the USA also promised another US$150 million for a 'forests for the future' initiative, an increase of US$217 million a year in overall environmental aid, and the Canadians promised an additional US$115 million a year. Then, the Japanese brought along the largest offer of new environmental aid, some US$500 million a year.

Some of the other promises appeared like public relations efforts to grab

favourable headlines that had little domestic backing. The EC promised an increase of US$800 million in environmental assistance at Rio, conditional on the approval of its twelve member governments. We have already quoted John Major as saying that he would back Agenda 21 financially. Promises like those of Canadian Prime Minister Brian Mulroney were also suspect. In Rio Mulroney announced a contribution of US$10 million to develop models of 'rational forest use in developing countries, based on Canadian principles'.[3] But Canadian NGO representatives at Rio like François Coutu of the UN Association in Ottawa say that 'rational forest use' in Canada consists of clearcutting huge areas of old growth forests, which is bitterly opposed by Canadian conservationists and indigenous peoples.[4]

At the same time the promises of new and additional money were threatened by recession. Months after Rio, the British Conservative government was planning to cut the following year's foreign aid budget by up to 270 million pounds sterling (US$437 million) or 25 per cent of its 1991 budget. And the EC was planning to slash its budget by £95 million (US$136 million). While governments insist that this new money will eventually come through, NGOs were expressing fears that the new money was simply replacing old aid. At the time many pointed out that despite the new programmes, as a rule aid money was becoming harder to get. Whether or not new bilateral money will actually be found remains to be seen. Let us now turn to the international institutional aspects of financial aid.

THE WORLD BANK

At the fourth PrepCom in New York, the G-77, following a lead from China (which is not a member of the group), wanted to create a new Green Fund to administer the financing of the Rio agreements. Each nation, rich or poor, donor or recipient, would have one vote on the Fund's activities. But donor countries refused to agree. As early as March, US President Bush voiced the opinion of most Northern leaders in saying that the GEF, the Global Environmental Facility, should be the 'primary vehicle' for any new money. Simultaneously, Michael Heseltine, the British trade and industry minister, relayed the same message from the British government.

The GEF had been established at a meeting in Paris in November 1990 to

'provide grants on concessional funding for technical assistance, preinvestment and investment projects and related activities in developing countries' in the fields of energy efficiency, foresting management, preservation of ecological diversity, water pollution, and protection of the ozone layer.[5] The GEF is a joint project of the World Bank and the two UN agencies, UNDP and UNEP. It is administered and controlled by the World Bank, which in turn is controlled by rich donor countries like the USA. The World Bank itself, which was merely an observer at the Summit process — it is not a UN agency although it is closely linked to the UN — was more than happy to take on this burden. Its president, Lewis Preston, told finance ministers at the Bank's 1992 spring meeting that developing countries would need more aid to meet the environmental goals agreed at the Summit. This aid, Preston said, should flow through development agencies 'with a proven track record in promoting development, reducing poverty, and protecting the environment' — meaning the Bank.[6] And at Rio the GEF was made the 'interim' financing agency for the biodiversity and climate change treaties signed in Rio, as it was already the financing agency of the so-called Multilateral Fund or Ozone Fund, which was established under the Montreal Protocol to protect the ozone layer. We present the GEF in more detail in the next chapter.

The World Bank is actually four agencies run by their member countries who put up capital to lend or guarantee loans to other member countries and the private sector. Its principal agency is the International Bank for Reconstruction and Development (IBRD), which was set up at a post Second World War meeting in July 1944 in Bretton Woods, New Hampshire, USA, to help finance the restoration of war-battered Europe. At this meeting two other institutions were planned under the auspices of the IBRD, namely the International Monetary Fund, which is supposed to help stabilize exchange rates, and the International Trade Organization (ITO) to set world trade rules. The latter was never set up, but an informal talking shop called the General Agreement on Tariffs and Trade (GATT) has taken its place and is still trying to formalize the ITO's existence. Collectively, the three are called the Bretton Woods institutions. The three other World Bank agencies are the International Finance Corporation, the International Development Agency (IDA), and the Multilateral Investment Guarantee Agency (MIGA). The International Finance Corporation was set up in 1956 and now has 145 members. It lends directly to the private sector and can even buy shares in companies. It also plays a role

in developing stock markets in Southern countries. Its 1991 loans totalled US$2.9 billion. The IDA was set up in 1960 and now has 140 members. It makes interest-free loans to the 'poorest' countries, which have to be paid back over 35 to 40 years after a 10-year period of grace. Its 1991 loans totalled US$6.3 billion for countries with a per capita income of US$1,195 or less. MIGA was set up in 1988 and has seventy-eight members. It spent US$922 million in 1991 to protect investors against nationalization or war, which commercial insurers refuse to cover.

According to its own literature, the IBRD — which had 156 member countries in early 1992 — now makes loans to Southern countries and the former Soviet bloc 'to help reduce poverty and to finance investments that contribute to economic growth. Investments include roads, power plants, schools, and irrigation networks, as well as activities like agricultural extension services, training for teachers and nutrition-improvement programs for children and pregnant women. Some World Bank loans finance changes in the structure of countries' economies to make them more stable, efficient and market-oriented. The World Bank also provides "technical assistance", or expert advice, to help governments make specific sectors of their economies more efficient and relevant to national development goals'.[7] Its loans totalled US$16.4 billion in 1991. The IBRD is therefore one of the biggest 'aid' donors to Southern countries alongside the 'aid' from the USA and Japan. Currently the biggest contributors are the USA (18.02 per cent), Japan (7.82 per cent), Germany (6.04 per cent), France and Britain (5.79 per cent each). Activists commonly call this system the 'one dollar, one vote' system. In addition, according to longstanding custom, the Bank's president is nominated by the USA and is a US citizen.

The activities of the World Bank group have been roundly condemned by development and environment activists. The major critics say that they are undemocratic because the donors have control while communities are not consulted about projects in their neighborhoods, that Bank loans are based on economic considerations alone ignoring other impacts such as environmental and cultural effects, that its plans and reports are not open to public scrutiny, and finally that the loans benefit the donor countries and rich élites in the Southern countries.[8]

There are problems even by the Bank's own standards. In mid-1992 a report was prepared by a special high level Portfolio Management Taskforce which

reported directly to the Bank's president. Willi Wapenhans, the chairman of the taskforce, submitted his draft recommendations after meeting with a number of policy-makers from borrowing countries and reviewing about 1,800 current Bank projects in 113 countries for which the Bank had lent US$138 billion. Entitled 'Effective Implementation: Key to Development Impact' and marked confidential, his report calculated that over a third of World Bank projects completed in 1991 were judged failures by the Bank's own staff, a dramatic 150 per cent rise in failures over the previous ten years.[9] Specifically the Wapenhans review noted that 37.5 per cent of the projects completed in 1991 were deemed failures, up from 15 per cent in 1981 and 30.5 per cent in 1989. Bank staff also said that 30 per cent of projects in their fourth or fifth year of implementation in 1991 had major problems. The worst affected sectors were water supply and sanitation, where 43 per cent of the projects were said to have major problems, and the agriculture sector with 42 per cent.

But let us look at two specific areas of World Bank financing – energy and forestry – as they are an integral part of the climate change and the biodiversity treaties signed at Rio and therefore likely to be managed by the GEF. A report prepared for the Summit by Greenpeace noted that the World Bank is the largest source of energy finance worldwide, lending billions of dollars for projects that increase the consumption of fossil fuels such as coal, gas, and oil that directly contribute to global warming.[10] Yet at the same time the Bank has no policy on global warming – its own 1991 reports indicated that it spent a mere 1 per cent of energy sector lending on energy efficiency, most of which was spent on studies. Interestingly, these studies showed that in Brazil and India as much as half of the projected new power demands could be met by energy efficiency and conservation practices. But what was the Bank itself doing? In the fiscal year ending in 1991, out of every five dollars that the Bank spent on energy, two were spent on gas and oil and eighty cents were spent on coal development. And in the five years leading up to 1996, the Bank was expected to spend over US$2.2 billion on gas and oil development and US$1.2 billion on coal development.[11]

Our second example is forestry. Between the time of its creation and 1992, the Bank financed eighty forestry projects worth over US$2.3 billion, on top of which it had paid for other projects such as road and dam construction which have also had a significant impact on forests. Many of these have resulted in significant deforestation and environmental damage. In 1978 the Bank began

to draft a forestry policy, which included what it called 'social forestry' which would benefit local communities and decrease deforestation. This policy, which was known as the Tropical Forest Action Plan (TFAP), was carried out with the help of the FAO and the UNDP. According to the FAO, the aim of these projects is to:

raise the standard of living of the rural dweller to involve him in the decision making process which affects his very existence and to transform him into a dynamic citizen capable of contributing to a wider range of activities than he was used to and of which he will be the direct beneficiary.[12]

In fact, according to Vandana Shiva of the Third World Network, who conducted an analysis of the TFAP published by the World Rainforest Movement, the Plan had exactly the opposite effect:

it takes forestry away from the control of communities and makes it a capital-intensive, externally controlled activity. It totally neglects the economics of tribal and peasant life based on natural forests and food production and focuses exclusively on the economics and production of commercial wood.[13]

Shiva makes three conclusions from her study of the TFAP. It fails to take into account the fact that international development financing is a cause of forest destruction, but rather puts the blame on the poor. As we have seen in the previous chapter, they do have a role, but only because development financing has forced them to that point. Second, she says that TFAP is based on the investment returns for commercial timber, and puts control in the hands of external commercial interests. Finally, it does not take into account the rights of indigenous peoples who have taken care of and depended on these forests for centuries. Because of the increasing opposition from activists and the obvious deforestation effects of TFAP and other Bank lending, the World Bank had its policy redrafted at the end of the 1980s by the World Resources Institute. But the new policy was still heavily criticized by activists.[14]

In short, the World Bank is clearly a development agency which has its origins in the Post-War ideology of unlimited economic growth. Growth is put before social and environmental costs. Later on, efforts are made to minimize these costs, rather than the Bank realizing that these effects may be symptoms of a larger problem with the original economic growth policies. Had the policy-makers looked at the root causes of poverty and environmental

degradation, such as the inequitable distribution of wealth, and the destruction of habitat and natural resources that poor people are dependent on, they might have realized that the environment and social issues are key factors, if not more important than economic growth. The question is whether an institution like the World Bank, whose primary mission was and still is economic growth and industrial development, is the most appropriate organization to manage the money that is supposed to lead to a more sustainable development. We, for our part, have serious doubts. Yet, beyond the institutional question, there is a profounder question of whether financial aid – be it multilateral or bilateral – is an appropriate answer to the global crisis to begin with.

AIDING THE NORTH

Previously, we have argued that focusing on money as a means of dealing with the global environmental crisis is at best inappropriate, as the problems are not of a technical or economic nature. Money is invested in further economic growth, which will only exacerbate the global environmental crisis. We would like to introduce here yet another argument, namely the fact that aid increases South–North inequity and thus promotes the exploitation and destruction of natural resources.

Indeed, it is often not mentioned that most aid, environmental or other, is repayable with interest. So that Uganda, for example, has to pay the World Bank back the money that it borrowed plus interest. Given that only a sixth of Bank projects in Uganda were actually successful because of the combined bungling of the Bank, aid agencies, and implementers, it is getting a mighty poor deal and little opportunity to find the money to pay back the loans.

At the end of the 1990s, the combined debt of poor countries was estimated at US$1 trillion. The net outflow of money from these countries to pay back this debt was quoted at US$39 billion in the year between 1 July 1987 and 30 June 1988. The World Bank estimates that it makes a profit of about US$1.1 billion a year (figures as of 1987), and this despite the fact that many countries did not pay back their debts.[15] Most of this money stems from the exploitation of natural resources, which goes hand in hand with environmental degradation and destruction. In the future we can expect financial outflows and thus environmental degradation to increase.

The purpose of most loans and of many grants is to generate profit for the donor country and its industries. This holds true even more for bilateral than for multilateral aid. Indeed, aid often pays for large projects such as the ones we have described above which largely sustain Northern construction contractors at the expense of the South's environment and poor communities. Studies of British and US aid show that these projects reap the most benefits not for the recipient country but for the donor country.[16] Vandana Shiva quotes J. Johnston, the US deputy assistant secretary of state in the Bureau of Economic and Business Affairs, as testifying to the US Congress in 1978 that for every dollar that his country paid into the multilateral development banks, $3 in business for US companies was generated.[17] Likewise, the British Overseas Development Administration has pointed out that for every pound that the government puts into the multilateral banks, it makes a 20 per cent profit.[18] For example, in 1985 Britain 'gave' Southern countries £531 million. But on the back of this 'aid' British companies won £637.2 million in contracts.[19]

Much of the profit generated from aid stems not simply from the lucrative construction contracts for building roads, dams and factories, but also from the commodities and labour exploited as a result of this new infrastructure. Japanese aid agencies, for example, pay to develop Malaysian fisheries because Japanese consumers want to buy fish, not to feed local people. Hong Kong companies invest in Sri Lankan textile factories because they want to profit from cheap labour. All this is part of the new free trade environment where goods are produced in the South for consumption in the North. This point was brought home to one of these authors when visiting the US embassy in San Salvador, El Salvador. The USAID deputy chief agreed to answer our questions. He was asked what return US taxpayers could expect for the US$1 billion in aid that their government had given to El Salvador. The official replied: 'The purpose of our aid is to get them to buy American products'.[20]

Why do Japanese aid agencies, for example, ignore the plight of local peoples? Because the aid is not for them, it is for Japanese industry and consumers, and it is to these they are listening. Japanese aid officers do not solicit the opinion of local peoples, let alone even spend much time at the project sites. Of the estimated 6,000 aid staff, fewer than 5 per cent are actually based overseas. Instead they are in Tokyo, where they are listening to Japanese corporations which are bidding for the contracts. And not surprisingly

this very same corporate sector played a significant role in encouraging governments to invest in more aid. In fact, the co-host of the 'eminent persons' Meeting on Financing Global Environment and Development convened in Tokyo by the Summit secretariat in April 1992 was none other than Keidanren, the federation of Japanese businesses, which had much to gain.

And having profited from construction and other development contracts handed over to Northern companies and from cheap labour, the North then profits again from the artificially low prices of Southern natural resources, largely due to the fact that the South is forced to sell its environmental assets in order to pay back the loans and interests that were given to it by the North in order to do all this. There is no reason and no argument why the mechanism we have described here will be any different when it comes to 'sustainable development' projects, as opposed simply to 'development' projects. The only thing that will be sustainable, we are afraid, is the very mechanism we have just described.

INSTITUTIONAL OUTCOMES

It can be argued that it is useful to have two institutions ensuring the follow-up of the UNCED process. One institution, the GEF, would deal with the financial aspects of the follow-up, while the other one, the Commission on Sustainable Development (CSD), would deal with the political aspects. Common sense would call for the political institution to oversee the financial one. However, it turns out that these institutions have no relationship. The GEF is an initiative of the World Bank, which not only wanted a piece of the UNCED cake, but moreover needed to address environmental concerns for internal reasons, attacked as it was by environmental activists. Establishing the GEF was its answer to this challenge. On the other hand, the CSD is the more logical institutional follow-up of UNCED. However, with the financial aspects being dealt with by GEF, the CSD has little power left. The official reason for this separation is that the GEF is dealing with the environmental conventions – climate change, bio-diversity, ozone convention, plus all future conventions – while the CSD is dealing with the follow-up of Agenda 21. In this chapter we critically analyse both these institutional outcomes of UNCED, the GEF and the CSD. Finally, we also mention the Earth Council, the post-UNCED vehicle for 'people power' as Maurice Strong likes to see it.

THE GLOBAL ENVIRONMENTAL FACILITY

In 1987 UNDP commissioned the World Resources Institute (WRI) to study the conclusions of the Brundtland Commission, which said that there was a

serious lack of funding for conservation projects and strategies which improve the resource base for development.[1] WRI came up with the idea of a financing facility to pay for environmental treaties. In September 1989, at the joint World Bank–IMF annual meeting (and in the wake of the Montreal Protocol on ozone depleting substances), the French government with backing from the Germans suggested that the Bank set up the Global Environmental Facility (GEF). The Bank invited UNDP and UNEP to a meeting with seventeen donor countries in Paris in March 1990, where an agreement was forged. Interestingly, the GEF was subsequently headed by one of WRI's senior analysts, Mohammed El-Ashry, who left WRI to take over the Bank's environment department.

The GEF's formal existence began in November 1990 as a three-year joint pilot project of the Bank, UNDP, and UNEP. Donor countries pledged US$861.4 million to the GEF's core fund, US$350.1 million in co-financing (Japan, Australia, Belgium, Canada, and Switzerland), and parallel financing. The existing US$200 million in the Montreal Protocol was included in the accounting, rounding off the initial pool at US$1.411 billion. GEF projects were invited in four areas – to tackle emissions of greenhouse gases, deal with ozone-destroying chemicals, protect biodiversity, or reduce pollution of international rivers. Projects had to be in countries with a per capita income of less than US$4,000 a year and the project had to deal with an international problem rather than merely with a local one. At the third meeting in April 1992 in Washington, DC, participants were asked to consider extending the four categories of schemes that then qualified for GEF money to cover other anticipated treaties. The new categories were combating the spread of deserts and tackling land-based forms of marine pollution, both of which were expected to and actually did feature quite strongly in initiatives proposed at the Earth Summit. By the fourth GEF meeting in December 1992 a total of seventy projects had been approved for US$584 million in financing. The GEF estimated at that point that 47 per cent of the core fund money had been earmarked for biodiversity, 36 per cent for global warming and 17 per cent for international waters.

In its next phase – the so-called GEF II – core funding will double or triple (US$2.8–4.2 billion), and the GEF will officially become the permanent funding mechanism for environmental conventions. This means that the GEF will have established itself as a new organization, despite the insistence of

mainly Northern countries at the fourth PrepCom in New York that no new institutions should result from UNCED. But what exactly is this institution about?

The emergence and the nature of the GEF is probably best understood in terms of organizational dynamics. Indeed, in the second half of the 1980s the World Bank came under heavy attack from environmentalists.[2] In response, the Bank created a new institution within itself, called the Environment Department. This department built itself up from two professional staff to 140 between 1987 and 1992, to the point where it had more environmental professionals than even UNEP itself. In fact, like the GEF that followed, the Environment Department was obviously built up to take over the new environmental agenda. Yet the new department was an afterthought, an addition to compensate for past failures, not to reset the agenda that created the problems in the first place. Even more than the Environment Department of the Bank, the GEF took advantage of the new opportunities that presented themselves because of the UNCED process. We have shed light on its emergence above.

But at the Earth Summit, the idea of using the GEF for post-Rio funding came under severe attack from Southern countries and NGOs because of the control wielded over it by the World Bank, against which they had been waging a long struggle. The way the GEF answered these challenges was by admitting to the various criticisms, without, however, fundamentally changing the nature of its activity, i.e. financing development projects. First of all, it made all its documents available to the public. Second, it consulted NGOs on each and every project, and even offered to pay for some of the NGOs' own projects. Finally, it offered Southern countries an equal say in its decision-making. And, of course, two other 'non NGO hostile' UN agencies were supposedly equal partners, so the World Bank would not control any of the decisions.

Let us start with the way the GEF handles Southern governments. When the GEF was set up, its agenda was dominated by the rich donor nations. Subsequently nine Southern countries – Brazil, China, Egypt, India, Indonesia, Mexico, Morocco, Pakistan, and Turkey – were invited to take part in the decision-making. They were also required to pay in US$5 million to be able to vote – half of this money could be provided by the Bank if the country did not have the ready cash. When it was pointed out that donor countries still outnumbered Southern countries after the second participants' meeting in

Geneva in December 1991, the number of participants was increased to thirty-two at the third meeting in April 1992 in Washington, DC, so half of the members were now Southern countries. Decisions were taken by consensus, and although there were no major disagreements up to that point, it was obvious that this would not last for too long. At its fourth meeting in Abidjan, Ivory Coast, the Bank reviewed a dozen proposals on voting structure. According to Ian Johnson, the GEF administrator at the Bank, they favoured one proposal that would allow for majority voting in the absence of a consensus, with the caveat that two-thirds of donor countries would have to vote on the motion. The most recent proposal is that there should be two tiers of voting. An administrative board of thirty members from thirty constituencies with geographic balance would vote the first time. The second vote would be weighed by dollars or both the North and the South would have veto power.

Though the GEF has moved away from the 'one dollar – one vote' principle of the Bank, the North will retain control. As a matter of fact, at the fifth meeting of the members of the GEF held in Beijing in May 1993, the proposals of the G-77 countries on the governance of the GEF funds were quite similar to those of the donor countries, probably because the South had received assurance on additional finance. Says Mohammed El-Ashry, the Director of the GEF: 'There was a new tone of cooperation at the meeting'.[3]

Not only Southern governments, but also NGOs are increasingly finding it easy to cooperate with the GEF in mutually productive ways. To begin with, the GEF managers had invited NGOs to the second participants' meeting in Geneva in December 1991. This meeting failed, even in the opinion of the GEF itself, because fewer than half of the participating countries showed up to meet representatives of forty-five NGOs, half of which were from the Southern countries. But at the third meeting in Washington, DC, the GEF extended personal invitations to (and offered to pay all of their costs) more NGO representatives including two of their strongest critics, Martin Khor and Vandana Shiva of the Third World Network. Unfortunately the meeting coincided with the big Southern countries' pre-Summit strategy meeting in Kuala Lumpur and neither of them was able to go.

In order to improve planning for this and future consultations with NGOs, the GEF set up a joint GEF/NGO taskforce in September 1992. In addition to the GEF's implementing agencies, the taskforce included four Washington-

based NGOs (WWF, IUCN, the Environmental Defense Fund, and the Bank's Information Center). However, recently the Committee broke up and only IUCN remains as a consulting agency with the GEF. NGOs were even offered the opportunity to implement or consult on GEF projects and, better still, the GEF offered to pay for some of these projects. This GEF/NGO cooperation, again, seems to profit the big Northern NGOs which will be the only ones to have a say in GEF. Already, according to Charles Abugre, a Ghanaian economist for the Accra-based Agency for Coordination and Development (ACORD) who also works for the TWN, WWF is the NGO most often consulted on GEF projects. As we have seen, there is no risk that the big Northern environmental NGOs' views will be radically different from the views of the GEF managers. As in the overall UNCED process, this process of establishing the GEF will have led to the further cooptation of the already quite mainstreamed NGOs (e.g. WWF, IUCN, and the Big 10), as well as to the fragmentation of the rest of the Green movement.

This also means that the GEF prevents itself from learning and transforming. Though it will take into account some concerns of Southern élites and big Northern environmental NGOs, the GEF will essentially be an agency which finances and manages projects related to environmental conventions. But they are development projects none the less. The Bank will exert major control, and it could well be that the function of the GEF turns out to be very similar to the function of its Environment Department, i.e. 'a Green add-on'. Indeed, despite the supposed equality of the three participating agencies – the Bank, UNDP, and UNEP – the actual day-to-day administration of the GEF itself is conducted by the World Bank. Both the chairman and the administrator are Bank employees, although theoretically the chairmanship is supposed to rotate among the agencies.

And there is also the matter of financing: in terms of project allocation, the Bank gets two-thirds of all the project money, UNDP gets the remaining third, while UNEP has no projects at all. Weighed by numbers of projects, fully two-thirds of the World Bank's projects are actually tied to its own investment projects of the kind we have criticized in the previous chapter, such as large dams, forestry projects, and the like.[4] According to estimates provided to us in April 1992 by Charles Feinstein, operations officer for the Bank's Global Environmental Unit, of twenty-three World Bank projects with US$298.5 million in approved funding in April 1992, there were seven free-standing

projects of a total of US$43.5 million, while the rest were tied to already existing projects – i.e. 70 per cent of the projects in number terms and 85 per cent of the money committed to Bank-related work. Also interesting is the fact that at least some GEF projects are designed to complement existing projects or 'greenwash' them. Vandana Shiva of the Third World Network gives the example of the Kerinci Seblat national park in Northern Sumatra, a GEF biodiversity conservation project which followed a World Bank investment project in forestry management programmes in 1991. Although a certain amount of documentation is available for the GEF projects, the usual Bank rules of non-disclosure apply to the massive companion Bank investment projects, which are often the source of the environmental problems.[5]

Not surprisingly, those NGOs which have been tracking the Bank for years are quite alarmed. Says Abugre: 'The World Bank, with its track record of environmental damage, cannot be entrusted with the role of the major institution for the management of the world's environment'.[6] An analysis of two GEF projects carried out by the Environmental Defense Fund and Greenpeace gives an even better picture of how the GEF is used to greenwash the World Bank's own activities and how the changes that the GEF is supposed to have brought to Bank thinking have not been implemented.[7]

EDF investigated a US$10 million GEF biodiversity protection project in the Congo called the Congo Wildlands Protection and Management Project, which the GEF described as free-standing. To begin with, EDF said that the project was not free-standing because it was tied to a US$20 million Natural Resources Management Project Bank loan that was being offered to increase Congolese timber exports. This Bank project also violated its own forestry policy which prohibits the logging of primary tropical forests. Finally, EDF pointed out that the root causes of deforestation in the Congo were landlessness and industrial expansion. These occur in the south of the country, not in the north where the project was located. The GEF project itself was paying for the construction of a road that would open up previously unvisited areas of the Nouabele Reserve to tourists and 'rational forest exploitation', but it could have the opposite effect, of bringing devastation to this untouched area.

Greenpeace investigated an US$8 million GEF grant to Costa Rica for Conservation and Sustainable Development of La Amistad and La Osa reserves on the Panamanian border and the Osa peninsula respectively. Greenpeace asked UNDP, which is conducting the project, for information on it and

received a total of five pages of information. Requests for more were denied. An NGO was identified for each of the two reserves, both of which were large foreign-funded groups (both were funded by the USAID). Only one – Fundácion Neotropica in La Osa – had received any documentation on the project. The other – Centro Cientifico Tropical – had received nothing. None of the groups listed by UNDP as project beneficiaries had any knowledge of the project and a government official said that they would be notified after the project had been designed. A Greenpeace survey in January 1992 of other NGOs working in these areas showed that none of them had been consulted by any of the groups conducting the project. Finally, no public meetings about the project had been held in any of the affected communities, despite the fact that the reserves covered areas that were home to 80 per cent of the country's indigenous peoples.

THE COMMISSION ON SUSTAINABLE DEVELOPMENT

The function of the Commission on Sustainable Development (CSD) is to promote and monitor the implementation of Agenda 21, in the various UN member states – the signatories of Agenda 21 – as well as in the numerous UN agencies. There are, in Agenda 21, about 2,500 recommendations in about 150 programme areas, ranging from poverty alleviation to toxic substances to their interrelationships.[8] Since the governments did not want any new institution emerging from UNCED, the CSD remains a commission that will meet two to three weeks annually and have a very small secretariat in New York. Its task is, of course, impossible.

Initially, the Commission was supposed to develop a plan to monitor the implementation of the overall environment and development activities of several of its own agencies, its member states and the multilateral development banks such as the World Bank. NGOs spent a lot of time campaigning for the setting up of such a commission and many viewed the Rio decision to create it as an important victory. William Pace of the Center for Development of International Law in Washington, DC, Tony Simpson, an Australian lawyer, Martin Khor of TWN, and Simone Bilderbeck from IUCN in the Netherlands, among others, spent a considerable amount of time convincing their

governments that such a commission would be a fitting forum to continue the debates initiated at the Summit. Unfortunately, the outcome is not exactly what they lobbied for: the Commission is bogged down in the UN bureaucracy, has no money and therefore no power, and grounds its activity on Agenda 21 which is toothless at best (see chapter 3). Furthermore, governments have asked the Commission to report to the Economic and Social Commission of the UN, a moribund and useless body which has little effect on policy.

Weeks after the Rio Summit, two teams of experts began to discuss possibilities for post-Rio activities. The first was composed of members of the Summit secretariat who had drafted the Summit agreements. Separately, UN Secretary-General Boutros-Ghali appointed the FAO chief, Edouard Saouma, to set up a taskforce to advise the secretariat on the new commission and also on how to coordinate environment and development activities within the UN system itself. In the eyes of many NGOs, Saouma was an unfortunate choice. In 1991, an open letter from eighty international NGOs, published in the *Ecologist*, had accused him of pushing the inappropriate industrialization of Southern agriculture and the export of cash crops instead of the production of food for local consumption. 'Whether in agriculture, in forestry, or in aquaculture,' the open letter claims, 'you have promoted policies which benefit the rich and powerful at the expense of the livelihoods of the poor. Policies that are, in effect, systematically creating the conditions for mass starvation'.[9]

In November 1992, one of the committees of the UN General Assembly set up an *ad hoc* open-ended working group on UNCED follow-up. The group met for three weeks, and drew up seven resolutions on various matters ranging from setting aside a day of the year dedicated to water [sic!], to plans for a special conference on migratory fish stocks, a conference on the sustainable development of small island developing states, and a convention on desertification, and plans to establish the CSD. At the next meeting of ECOSOC, in early February 1993, governments continued to flesh out these plans and agreed to have a special session at the end of the month to organize the workplan of the CSD for the next few years. As soon as the ECOSOC meeting ended, the governments sat down to elect the members of the CSD. Some eighty states took part in the elections for the fifty-three seats on the Commission. Under a system aimed at ensuring geographical equity, thirteen members were elected from Africa, thirteen from Western Europe and North America,

eleven from Asia, ten from Latin America, and six from Eastern Europe.

Shortly thereafter the CSD held a first organizational session in New York. At that time, Secretary-General Boutros-Ghali suggested that the forty chapters of Agenda 21 should be divided into nine discussion clusters. (The first chapter, an introduction, is not included.) They were:

(a) Critical elements of sustainability (chapters 2–5);
(b) Financial resources and mechanisms (chapter 33);
(c) Education, science and technology (chapters 16, 34–37);
(d) Decision-making structures (chapters 8, 38–40);
(e) Roles of major groups (chapters 23–32);
(f) Health and human settlements (chapters 6, 7 and 21);
(g) Land, forests and biodiversity (chapters 10–15);
(h) Atmosphere, oceans and fresh water (chapters 9, 17–18);
(i) Toxic chemicals and hazardous wastes (chapters 19, 20 and 22).

The first five clusters were to be taken up every year at the annual 2–3 week meetings of the CSD. For the first session of the CSD the governments were also supposed to discuss financial commitments and the transfer of environmentally sound technologies. In 1994, the governments are supposed to give special consideration to international cooperation and changing consumption patterns (from cluster a), health and human settlements (cluster f), and toxic chemicals and hazardous wastes (cluster i). The year after, governments are supposed to discuss the chapters on combating poverty and demographic patterns (from cluster a) and land, forests and biodiversity (cluster g). In 1996 the CSD are supposed to take up cluster h.

The governments also elected Malaysian Ambassador Razali Ismail as chairman of the CSD. All other officers also came from the UNCED negotiations. The CSD held its first substantive meeting in New York in June 1993. Inter-sessional groups on technology transfer and finance were established, as well as an Inter-Agency Committee on Sustainable Development, a group of nine UN agency representatives chaired by Nitin Desai, another UNCED person, who has meanwhile advanced to the position of Under-Secretary-General of a new Department of Policy Coordination and Sustainable Development. It was also requested that numerous reports be prepared by the secretariat for the next year's meeting. This includes a report

on the progress made in the implementation of Agenda 21.

As for outside input into the CSD, we detect a set-up that is identical to the UNCED process: on the one hand, NGOs will have to go through complex procedures in order to feed into the CSD, while, on the other hand, business and industry and other mainstream voices will have a direct line. To recall, at the Summit, governments did agree to allow NGOs to play an 'expanded' role on the Commission. The secretariat was looking at two options: either having representation from certain major NGOs and coalitions or asking the NGOs to organize themselves into constituencies and propose representatives. Razali, who met with several NGOs on an *ad hoc* basis in spring 1993, said he would like to see NGOs reporting to special committees in the Commission on matters on which they have expertise. But he underlined the fact that the Commission was to be a body of governments and cautioned that NGOs would not replace governments as the voice of the people, but simply give expert opinions when necessary.

On the other hand, in spring 1993 UN Secretary-General Boutros-Ghali set up a high level Advisory Board on Sustainable Development to contribute to the work of the CSD. The Board will input directly into the CSD and ECOSOC, as well as the UN's Administrative Committee on Coordination, the body which coordinates the activities of the UN's different agencies. Among the thirteen members of the Board, all chosen of course for their 'knowledge of environment and development',[10] we find, not surprisingly, the by-now established environment and development élite, such as Klaus Schwab of the Davos Forum, Stephan Schmidheiny, and Maurice Strong.

THE EARTH COUNCIL

During the Summit Maurice Strong proposed setting up a group of independent experts to be called the Earth Council and to be based in Costa Rica. He hoped that the governments would mandate this body to monitor the results of the Summit. Strong saw this as a continuation or follow-up of the Global Forum. When his proposal was largely ignored by governments, except for the Central American governments, he went back to EcoFund and the private sector to get more backing, this time for an 'NGO monitoring group'. In September, when Strong's job as secretary-general of the Summit was

terminated, he became the chairman of the organizing committee for the Earth Council. Also on this organizing committee was Stephan Schmidheiny and Benjamin Read from EcoFund (ex-officio). At that point, EcoFund received the bulk of the Swatch contribution to the EcoFund (US$1.8 million) and offered an unstated share of it to the new Earth Council, which has other strong financial backers. It is also being co-sponsored by the IUCN, ICSU and the Society for International Development (SID), and has been endorsed by the WRI. In October 1992 Strong and President Rafael Calderon of Costa Rica inaugurated the Earth Council in San José, the country's capital, to 'promote worldwide awareness, understanding and resolution of major planetary problems'.[11]

But since then the Earth Council has not really taken off, and its Geneva subsidiary has closed down for lack of funds. Also, environmental activists like Greenpeace sceptically pointed out that the Council might just be a front-office for business to promote its own agenda. Josh Karliner, coordinator for Greenpeace's Earth Summit activities, said:

The Earth Council has some questionable sponsors like Swiss billionaire, Stephan Schmidheiny, whose Business Council for Sustainable Development used the Earth Summit to raise multinational corporations to the status of global environment leaders. But what they do in actual practice is very different and Greenpeace is concerned that the Earth Council may serve to institutionalize the greenwash of corporations promoted by Schmidheiny and Strong in Rio.[12]

In a recent article promoting the Earth Council as a vehicle for 'People Power' Strong does not deny this. There seems some confusion about the distinction between grassroots and business. For example, he wrote in spring 1993:

There has been a dramatic rise in Rio-induced activities and initiatives at the grass-roots level – and what I call the 'brass-roots' level: the influential constituencies of business, financial, scientific and professional organizations [sic!]. We must count on the energies generated by these activities to keep the follow-up and implementation of the agreements reached at Rio at the center of the political agenda. The Earth Council, now in the process of formation, is designed to facilitate this.[13]

CONCLUSION: WAS IT WORTH THE EFFORT?

When the Rio Conference ended, optimists argued that the awareness of international leaders had been raised and the leaders were now firmly committed to the principles of 'sustainable development'. What's more, they pointed to the fact that some new money and institutions were now being planned. One year after the Rio Conference, we can say that even the most modest hopes have been disappointed. The institutional and financial outcomes bear no relationship to the expectations raised, let alone to the needs and the urgency of the global environmental crisis. Moreover, the financial and the institutional outcomes have to be divided, thus reducing the chances even further that the 'means', i.e. the money, will be related to the theoretically more substantial 'goals'.

Of course, it would be inappropriate for us to complain about these outcomes, as we have argued all along that the philosophy of sustainable development on which UNCED was based and the set-up of its process were flawed from their inception. However, one could have hoped that along the way some learning would occur among the various agents and stakeholders involved. One could have hoped that such learning, had it occurred, would translate into innovative institutional structures which would at least carry this learning process further, and perhaps initiate a collective environmental learning dynamic of its own. The opposite has been the case: because of its set-up as a lobbying and as a public relations exercise, UNCED promoted established worldviews and cemented existing institutions. Except for some lofty 'global environmental awareness raising', collective learning does not seem to have occurred.

Instead, old thinking about economic growth prevails, old institutions promoting such growth persist, and the old development establishment that had made a living out of such economic growth has repackaged itself in green and miraculously represented itself as the new global environmental leaders. The monitoring of what is believed to be the solution to the global crisis is handed over to a powerless commission, more interested in sustaining the UN and the nation-state system than in questioning whether the process it is monitoring is actually leading somewhere. On a totally separate track, the implementation of environment and development projects has been appropriated by a subsidiary of

the biggest industrial development and economic growth agency, the World Bank. If the GEF does not directly sustain the development paradigm – though in a subtle way it does precisely that – it will at least serve as a green smokescreen, barely hiding the fact that behind it the development paradigm has, once again, survived. If the Bank now has its Green facility, the UN a Green commission, and business and industry two or three Green councils, the Green movement after Rio still has to apply for accreditation. It has to, and most likely will, beg for participation in a process which will further weaken it.

Part V

CONCLUSIONS

11

WHAT NOW?

At the Summit, the day before George Bush was to address the assembled
government representatives and heads of state, youth representatives were
given a few minutes to sum up their views of the process. The various youth
groups that were present elected Wagaki Mwangi from Kenya, a representative
of the Nairobi-based International Youth Environment and Development
Network, to speak for them.

She said:

[The Summit] has attempted to involve otherwise powerless people of society in the process.
But by observing the process we now know how undemocratic and untransparent the UN
system is.

At this point the closed circuit TV transmission of the speech to observers
outside the hall was cut, apparently caused by a technical problem, but many
of the youth felt that it was deliberate censorship.

Unheard by all except the delegates inside the hall, Mwangi continued.

Those of us who have watched the process have said that UNCED has failed. As youth we beg
to differ. Multinational corporations, the United States, Japan, the World Bank, The
International Monetary Fund have got away with what they always wanted, carving out a better
and more comfortable future for themselves . . . UNCED has ensured increased domination by
those who already have power. Worse still it has robbed the poor of the little power they had.
It has made them victims of a market economy that has thus far threatened our planet. Amidst
elaborate cocktails, travelling and partying, few negotiators realised how critical their decisions
are to our generation. By failing to address such fundamental issues as militarism, regulation of
transnational corporations, democratisation of the international aid agencies and inequitable
terms of trade, my generation has been damned.[1]

SUMMARY

The decision to have the Earth Summit was an outcome of the UN mandated World Commission on Environment and Development led by Norwegian premier Gro Harlem Brundtland which reported back to the UN on the state of the planet in 1987.

The report of the Commission, as we have shown, avoids the issue of the ownership of resources. According to it, the global commons consist of Antarctica, the high seas, and outer space. Land and biological resources such as seeds are not considered part of the commons. It says that the main cause of environmental degradation is population and poverty – too many people with low living standards who are destroying the environment. Little mention is made of the problems of over-consumption – that most of the resource depletion is caused by a few people consuming too much – or the fact that poverty is caused by export orientation in the face of falling commodity prices. Although some importance is given to indigenous people and women, communities and local organisations are given short shrift.

The Commission was created at the height of the Cold War in 1983 when NATO missiles were being implanted in Western Europe. After a decade and a half of standstill or even deterioration in global cooperation, the environment appeared to offer a way forward that avoided the traditional East–West deadlock. However, it was rapidly rediscovered – as this had already been found at the 1972 Stockholm Conference on the Human Environment – that addressing the environment from a resource management perspective opened up a potential South–North conflict. If military build-up was a relatively easy way to address the East–West confrontation, the North–South environmental conflict was less easy to tackle.

The Brundtland Commission found another way around the problem. It promoted the idea of sustainable development, similar to what had been suggested in 1972. This defined an efficient natural resource management approach within the broader context of industrial development coveted by both Northern and Southern governments. Here sustainability is defined in terms of the resource base, not in terms of society, culture, and people. It basically talks of sustainable growth, not of sustainable communities. But the Brundtland report did more than simply advocate sustainability; it was a staunch advocate of growth as the principal means to stop the poor from destroying the environment.

Meanwhile, with the end of the Cold War, a new factor had entered the picture: global ecology and its new approach to environmental problems. Global ecology – exemplified in a satellite picture of the fragile planet Earth taken from space – was a threat to nation-states and national politics far more deep seated than any previous social and environmental movements, because it fundamentally questioned their and their sovereignty's relevance as units when dealing with the global ecological crisis. If nation-states are perhaps pertinent agents to promote industrial development, a way out of the global crisis caused by such development will have to be found simultaneously below and above the nation-state level, i.e. in the interaction between the local communities and the global awareness. Global ecology also posed a threat to business and industry because it raised the issue of the ultimate limits to growth on a finite planet far more acutely than the environmental movement did in the 1970s.

The Summit came right on time to defuse both these threats from global ecology. To do so, the Brundtland Commission, and later the UNCED secretariat used two political tools from the Cold War. Both became particularly helpful in the overall transformation of the global ecological crisis into global environmental management, while putting adversarial environmental movements to work for them. These are the security analogy and the New Age model of politics.

The security analogy originates in the debate for security from nuclear weapons in the beginning of the 1980s, a debate which can be traced back to the origins of the atomic bomb. Environmental degradation, together with problems of development such as poverty, are said to be a threat to the security of humanity, so humanity has to combat this threat by mobilizing all available means to exterminate it. This analogy automatically leads to a resource, a risk, and ultimately to a crisis management approach, where the most efficient way to deal with the crisis will be a militaristic one, based on high-tech and hierarchy.

Parallel to and in support of this view, the New Age model of politics says that since we are all faced with an unprecedented threat and are equally endangered, we must all join hands as humans in order to overcome the threat. We have to, it is argued, work together for a common purpose. The more powerful among us will, quite logically, have to take the upper hand to lead the process. It is in the North and through facing peace and security issues

during the Cold War that this New Age model of politics has emerged. The model stresses individual awareness of the threat and is supposed to lead to collaboration and cooperation. The UNCED secretariat and other self-appointed organizations such as the Center for Our Common Future helped organize this 'cooperation'.

Both the security analogy and the New Age model of politics were the basis for organizing the so-called 'independent sectors' at the Summit. The secretariat accredited every single NGO that applied (except four), a move which simultaneously strengthened the secretariat because the more NGOs the secretariat could line up, the more legitimacy it had and the stronger it became to overcome any dissenting opinions from either governments or NGOs. NGO coalitions also played the game. They organized NGOs to speak with one voice, applying the rationale that since we are all in the same boat, everybody should make his or her contribution to global management. And in order to achieve this, money was willingly provided by business and foundations. As a result of this 'facilitating' process, business and industry, which do share a common culture and working methods, came out strengthened, while the culturally diverse environment and development movement diluted its inherent strength stemming from its very diversity and unique approach to local situations. Indeed, this New Age model of politics turned out to be nothing more than the lobbying model of US democracy. The model has, not surprisingly, strengthened the rich and powerful lobbyists.

Interest from Southern governments picked up as the Summit PrepComs got under way. The South Commission, which was closely modelled on the Brundtland Commission but composed entirely of representatives from Southern countries and far more focused on the issues of development, put out its report. Although not a part of the official Summit process, it summarized the thinking and cooperation between Southern governments at the time. Like Brundtland, the report does not consider the common ownership of resources and it too advocates speeding up growth to solve the problems of poverty. It focuses largely on cooperation between nation-states and the use of trade for this purpose. And at the Summit, the Southern governments discovered and picked up on NGO demands such as the equitable distribution of profits as the key leverage to justify further industrial growth for the South. If the North–South deadlock became portrayed as the ultimate global crisis, this was only in order to allow further industrial development miraculously to emerge as the only solution to it.

Business and industry, in particular big business, were systematically built up through the UNCED process as the agents holding the key to solving the global ecological crisis. Since North and South came to agree that accelerated economic growth was the solution, TNCs had no trouble presenting themselves as the agents which could further stimulate such growth, provided, however, that environmentally based trade restrictions would not impede them. Under the influence of some new management philosophies and helped by public relations, big business proposed the only intellectual novelty of the official UNCED process, i.e. 'clean' growth, 'clean' meaning technological and organizational efficiency. Not only did this eco-efficiency approach become widely accepted, but big business managed, thanks to its privileged access and its generous financial contributions, never even to be mentioned in the UNCED documents as being a problem for the environment, locally and globally. Two other major contributors to the global crisis, science and the military, also managed never to be mentioned as a problem.

At Rio itself, the agreements reflected the ideologies that had gone into their creation. Neither Brundtland, nor the secretariat, nor the governments drafted plans to examine the pitfalls of free trade and industrial development. Instead they wrote up a convention on how to 'develop' the use of biodiversity through patents and biotechnology. Likewise the governments drafted a statement of principles on the protection of forests that says nothing about logging, one of the major threats facing them. They also drafted a convention on climate change that does not even have a concrete deadline or targets by which to achieve its aims. The two other documents – Agenda 21, an 800-page action plan for the planet to achieve sustainability by the twenty-first century, and the Rio Declaration of Environment and Development, a set of twenty-seven rights and duties of peoples and states towards the planet – remain lofty and toothless.

Of course, the relevance of nation-states as units for dealing with the crisis is not questioned, since the various national governments are in charge of implementing these agreements. If the local peoples are mentioned, this is to ensure their participation in the national resources management and development plans.

What is more, the institutional outcomes are not even up to the quite limited challenges as UNCED has defined them. The Global Environmental Facility (GEF) will distribute money to specific projects related to the

conventions. Because of the largely predominant eco-efficiency approach and because the GEF remains under the World Bank's shadow, it is likely that these will be development projects in the very tradition of Northern development aid. The GEF is indeed the vehicle of the new global managers, who believe in global technocratic solutions, see environmental problems as a threat to human security, and seek to solve them with either a resource or a risk management approach. On the other hand, the Commission on Sustainable Development, the other institutional outcome of the Summit, will be a body of politicians and turn into yet another UN talking shop with no decision-making power. NGOs will be allowed to lobby both, but their influence is likely to be as limited as it was in the general UNCED process.

Overall, UNCED did not offer any vision or way out. There was no alternative to the still dominant development paradigm, not even a critique of it. No thought had been given to the process at all, and no major stakeholders – not even the secretariat and the Secretary-General – seem to have understood what was at stake in Rio. And this is what Maurice Strong and his secretariat ultimately must be criticized for in our view, i.e. the absence of solid intellectual leadership or serious vision. As a result, the development paradigm was given, through UNCED, yet another period of grace.

BALANCE SHEET AND CHALLENGES

This book has offered a critique of the UNCED process and a critical look at the various aspects of development that were being played out through UNCED. It has concluded that as a result of this process the planet and most of its inhabitants will be worse and not better off. After UNCED, just as before, we do not have any answer to the increasingly pressing global environment and development crisis, not even to aspects of it.

The major lesson to be drawn from the entire ten-year process leading up to UNCED is, in our view, that the global approach is at best a useful tool for awareness raising. But it is not at this global level that the environment and development crisis will be dealt with. Overall, the shift to the global approach, as it has occurred through UNCED, seems only to have reproduced the old approaches and solutions, this time on a planetary scale. Rather than facing up to the challenge of the limits to growth and the prospect of deindustrialization,

UNCED has raised the promotion of economic growth to a planetary imperative. Rather than developing a new vision in line with the challenges of global ecology, UNCED has rehabilitated technological progress and other cults of efficiency. Rather than coming up with creative views on global governance, UNCED has rehabilitated the development institutions and organizations as legitimate agents to deal with new global challenges. These include the Bretton Woods institutions and the UN, as well as the national governments and the multinational corporations. And, finally, rather than making the various stakeholders collaborate and collectively learn our way out of the global crisis, UNCED has coopted some, divided and destroyed others, and promoted the ones who had the money to take advantage of this combined public relations and lobbying exercise.

As a matter of fact, this shift to the global, which UNCED has significantly helped to achieve, turns out to be the continuation of the development process, the logical outcome of the pursuit of economic growth through techno-scientific progress, this time on a planetary scale. Yet more economic growth, better technologies, more efficiency and increasing planetary management will at best help us buy some time.

We think that the only way out of this crisis is to question this development process in its entirety. Given that the biosphere is a closed system, we must come to admit that the system cannot grow to the point when it will develop sustainability. We must acknowledge that industrial development has induced global cultural and ecological changes of an unprecedented nature which will further restrict, not increase, our possibilities within that system. And we must accept that further industrial development will only lead to further destruction. Instead, we must think and collectively behave in terms of the sustainability of a closed and finite system of local and regional resources, as well as of socially and culturally rooted users.

UNCED has shown us the global horizon, but by analysing the UNCED process we now know that the word 'global' is a mirage. It turns out to be the illusion created by the traditional agents and major stakeholders in order to maintain their privileges and to avoid questioning the fact that their traditional problem-solving mechanisms are basically bankrupt. If the global perspective, as UNCED shows, is the increasingly blocked horizon and if global management — total quality or not — is a dead end, we have no choice but to focus on the local, its people, and its communities . . . and collectively un-learn the development paradigm of which modern society is both the product and the victim.

NOTES

INTRODUCTION

1. L.J. Goree VI, 'Summiteers, sherpas and the rest', *Crosscurrents*, 26 March 1991, p. 9.

2. *EcoCurrents*, Vol. 2, No. 2, May 1992. (Obtainable from: 1 Emile-Nicolet, 1205 Geneva, Switzerland.)

3. W. Sachs (ed.), *The Development Dictionary*, London, Zed Books, 1991, p. 2.

1 WHOSE COMMON FUTURE?

1. World Commission on Environment and Development, *Our Common Future*, Oxford, Oxford University Press, 1987, p. 1.

2. Like the Brundtland Commission on Environment and Development, the so-called Palme or World Commission on International Disarmament and Security was addressing, at the beginning of the 1980s, the question of the threats to humanity's security in the light of the New Cold War. See *Common Security: A Blueprint for Survival*, New York, Simon & Schuster, 1982.

3. The World Commission, op.cit., p. 304.

4. The World Commission, op.cit., p. X.

5. The World Commission, op.cit., p. 8.

6. The World Commission, op.cit., p. 95.

7. The World Commission, op.cit., p. 95.

8. The World Commission, op.cit., p. 105.

9. The World Commission, op.cit., p. 96.

10. The World Commission, op.cit., p. 98.

11. The World Commission, op.cit., p. xx.

12. E.F. Schumacher, *Small is Beautiful. Economics as if People Mattered*, London, Blond & Briggs, 1973.

13. World Bank, *World Development Report 1992: Development and the Environment*, Oxford, Oxford University Press, 1992, p. 11.

14. The World Commission, op.cit., p. 12.

15. The World Commission, op.cit., p. 123.

16. The World Commission, op.cit., p. 125.

17. C. Fowler and P. Mooney, *Shattering Food Politics and the Loss of Genetic Diversity*, Tucson, Arizona, Arizona University Press, 1990, p. 70.

18. The World Commission, op.cit., p. 144.
19. The World Commission, op.cit., p. 144.
20. The World Commission, op.cit., p. 150.
21. The World Commission, op.cit., p. 150.
22. The World Commission, op.cit., p. 13.
23. The World Commission, op.cit., p. 147.
24. The World Commission, op.cit., p. 13.
25. The World Commission, op.cit., p. 159.
26. The World Commission, op.cit., p. 206.
27. The World Commission, op.cit., p. 173.
28. The World Commission, op.cit., p. 43.
29. The World Commission, op.cit., p. 196.
30. The World Commission, op.cit., p. 196.
31. The World Commission on International Disarmament and Security, *Common Security*, op.cit.
32. The World Commission, op.cit., p. X.
33. The World Commission, op.cit., p. 304.
34. The World Commission, op.cit., pp. 302–303.
35. The World Commission, op.cit., pp. 297–298.
36. The World Commission, op.cit., p. 301.
37. The World Commission, op.cit., p. 301.
38. The World Commission, op.cit., p. 302.
39. For a clarification of the meaning of the term 'commons' see: E. Ostrom, *Governing the Commons. The Evolution of Institutions for Collective Action*, Cambridge, Cambridge University Press, 1990.
40. The World Commission, op.cit., p. 8.

2 SOUTHERN ELITES

1. The South Commission, *The Challenge to the South*, Oxford, Oxford University Press, 1990, p. 61.
2. The South Commission, *The Challenge to the South*, op.cit., p. 58.
3. The South Commission, *The Challenge to the South*, op.cit., p. 59.
4. The South Commission, *The Challenge to the South*, op.cit., p. 21.
5. The South Commission, *Environment and Development. Towards a Common Strategy of the South in the UNCED Negotiations and Beyond*, Geneva, The South Center, 1990.
6. The South Commission, *The Challenge to the South*, op.cit., p. 113.
7. The South Commission, *The Challenge to the South*, op.cit., p. 86.
8. The South Commission, *The Challenge to the South*, op.cit., p. 129.
9. The South Commission, *The Challenge to the South*, op.cit., p. 180.
10. The South Commission, *The Challenge to the South*, op.cit., p. 131.
11. The South Commission, *The Challenge to the South*, op.cit., p. 132.
12. The South Commission, *The Challenge to the South*, op.cit., p. 132.
13. The South Commission, *The Challenge to the South*, op.cit., p. 81.
14. The South Commission, *The Challenge to the South*, op.cit., p. 81.
15. The South Commission, *The Challenge to the South*, op.cit., p. 93.
16. The South Commission, *The Challenge to the South*, op.cit., p. 87.
17. The South Commission, *The Challenge to the South*, op.cit., p. 138.
18. The South Commission, *The Challenge to the South*, op.cit., p. 138.
19. The South Commission, *The Challenge to the South*, op.cit., p. 141.

20. The South Commission, *Environment and Development*, op.cit., p. ii.

21. The South Commission, *Environment and Development*, op.cit., p. iii.

22. The South Commission, *The Challenge to the South*, op.cit., pp. 91–92.

3   RIO AND BUST

1. *Third World Resurgence*, 1992, No. 24/5, p. 12.

2. P. McCully, draft article for *Media Natura*, London, 1992.

3. V. Shiva, *Third World Network Briefings*, 1992, Paper No. 24, Third World Network, Penang, Malaysia, p. 80.

4. *Third World Resurgence*, 1992, 24/25, p. 16.

5. Convention on Biological Diversity, 5 June 1992, article 3. (Obtainable from: UN Department of Public Information, attn. Project Manager for Sustainable Development, Room S-894, United Nations Plaza, New York, NY 10017.)

6. Article 16(1), Convention on Biological Diversity, op.cit.

7. D. Sitarz (ed.), *Agenda 21: The Earth Summit Strategy to Save Our Planet*, Boulder, Colorado, Earth Press, 1993, p. 119. This is an edited version of Agenda 21. We refer here to this version rather than to the original 800-page text.

8. *Third World Resurgence*, 1992, 24/25, pp. 19–20.

9. *Tiempo*, 1992, No. 6, p. 1.

10. Calculated from World Bank statistics in: *The World Development Report 1992: Development and the Environment*, Oxford, Oxford University Press, 1992, p. 204.

11. *Policy Implications of Greenhouse Warming*, Washington, DC, National Academy Press, 1992.

12. *Tiempo*, 1992, No. 6, p. 1.

13. Statement on Forest Principles, Official Document. (Details as for note 5.)

14. *Third World Resurgence*, 1992, 24/25, p. 12.

15. Statement on Forest Principles, Paragraph 2(a). (Details as for note 5.)

16. Sitarz, op.cit., 1993, p. 94.

17. *Network '92*, 1991, XI, p. 13.

18. Greenpeace forest activist Francesco Martone quoted in: M. McCoy and P. McCully, *The Road from Rio*, Amsterdam, World Information Service on Energy and International Books, 1993, p. 94.

19. *Earth Island Journal*, 1993, Winter, p. 18.

20. See: Sitarz, op.cit., 1993; M. Keating, *The Earth Summit's Agenda for Change. A Plain Language Version of Agenda 21*, Geneva, The Center for Our Common Future, 1993.

21. Sitarz, op.cit., 1993.

22. This critique will be addressed in Part III.

23. Sitarz, op.cit., 1993, p. 10.

24. Sitarz, op.cit., 1993, p. 19.

25. Greenpeace International, *UNCED Undone: Key Issues Agenda 21 Does Not Address*, Amsterdam, Greenpeace, 1992.

26. Financing of sustainable development will be discussed in Part IV.

27. Sitarz, op.cit., 1993, p. 21.

28. D. Meadows *et al.*, *The Limits to Growth*, New York, Universe, 1972.

29. *Earth Island Journal*, 1993, Winter, p. 18.

4 TELLING 'GREENS' APART

1. See: M. Finger (ed.), *The Green Movement – Worldwide*, New Hartford, CT, JAI Press, 1992.

2. See for example: *Society and Natural Resources*, 1991, Vol. 4, No. 3, special issue, 'Two decades of American environmentalism', guest editors: R. Dunlap and A. Mertig.

3. IUCN/UNEP/WWF, *The World Conservation Strategy*, Gland, Switzerland, IUCN, 1980.

4. M. Dowie, 'The New Face of Environmentalism', *UTNE Reader*, July/August 1992, p. 107.

5. F. Pearce, *Green Warriors*, London, Bodley Head, 1991, p. 84.

6. Dowie, op.cit., p. 108 and p. 110.

7. P. Watson, *Earth First! Journal*, 21 December 1986, p. 1.

8. See: M. Finger, 'UNCED – an exercise in New Age politics', in W. Sachs (ed.), *Conflicts in Global Ecology*, London, Zed Books, 1993.

9. See: D. Korten, *Getting to the 21st Century. Voluntary Action and the Global Agenda*, West Hartford, CT, Kumarian Press, 1990.

10. D. Korten, 'Third Generation NGO Strategies: a Key to People Centered Development', *World Development*, 1987, Vol. 15, Supplement, p. 148.

11. Korten, 1990, op.cit.

12. Korten, 1987, op.cit., p. 149.

13. V. Shiva and D. Bandyopadhyay, 'The Chipko Movement', in J. Ives and D. Pitt (eds), *Deforestation: Social Dynamics in Watersheds and Mountain Ecosystems*, London, Routledge, 1988, pp. 224–241.

14. See: *The Ecologist*, Special issues 'Whose Common Future?', July/August 1992.

5 FEEDING THE PEOPLES INTO THE GREEN MACHINE

1. United Nations General Assembly Resolution 44/228, p. 9.

2. A/CONF/151/PC/2. (Details as for note 5, Chapter 3.)

3. *Brundtland Bulletin*, June 1990, No. 8, p. 1.

4. K. Dawkins, 'Sharing rights and responsibilities for the environment. Assessing potential roles for non-governmental organizations in international decision-making', unpublished master's thesis, Cambridge, Massachusetts, M.I.T., 1991, p. 46.

5. See: A/45/46, p.22. (Details as for note 5, Chapter 3.)

6. W. Lindner interviewed by Stephen Lerner. See: S. Lerner (ed.), *Earth Summit. Conversations with Architects of an Ecologically Sustainable Future*, Bolinas, California, Common Knowledge Press, 1991, p. 241.

7. Lindner in Lerner, op.cit., 1991, p. 242.

8. Center for Our Common Future, internal paper, undated.

9. *Network '92*, December 1990, p. 5.

6 WHAT DID ENVIRONMENTAL NGOs ACHIEVE?

1. A. Doherty, *NGOs and UNCED*, excerpted with permission from a draft report for the International Institute for Applied Systems Analysis, Laxenburg, Austria, 1992.

2. 'Editorial', *Crosscurrents*, CONGO, Geneva, 30 March 1992, p. 2.

3. M. Valentine, *Earth Island Journal*, 1992, Summer, p. 40.

4. This information has been provided to us by Kristin Dawkins from the Institute on Agriculture and Trade Policy, Minneapolis, USA.

5. K. Dawkins, 'Sharing rights and responsibilities for the environment. Assessing potential roles for non-governmental organizations in international decision-making', unpublished master's thesis, Cambridge, Massachusetts, M.I.T., 1991, p. 49.

6. Unpublished survey conducted by Ann Doherty, International Institute for Applied Systems Analysis, Laxenburg, Austria, July 1992.

7. M. Khor, *Third World Resurgence*, No. 24/25, p. 4.

8. *Financial Times*, 15 June 1992, p. 7.

9. F. Pearce, *New Scientist*, London, 27 June 1992, p. 12.

10. M. Finger and P. Chatterjee, *EcoCurrents*, Vol. 2, No. 2, May 1992, p. 2.

Part III

1. United Nations Commission on Transnational Corporations (UNCTC), *Ongoing and Future Research: Transnational Corporations and Issues Relating to the Environment*, New York, April 1989, p. 6.

2. UNCTC, *Transnational Corporations and Issues Relating to the Environment. The Contribution of the Commission and UNCTC to the Preparatory Committee for UNCED*, New York, February 1991; UNCTC, *Ongoing and Future Research*, op.cit.; A. Makhijani *et al.*, *Climate Change and Transnational Corporations. Analysis and Trends*, New York, UNCTC, 1992, Environment Series No. 2, p. 77.

7 PROMOTING BIG BUSINESS AT RIO

1. M. Wald, 'GM signs accord with environmentalists', *New York Times*, 9 July 1992, p. C4.

2. *Brundtland Bulletin*, Sept/Dec 1990, SF18.

3. *Brundtland Bulletin*, March 1991, p. 69.

4. *Network '92*, No. 6, 1991, p. 15.

5. *Network '92*, October 1990, p. 11.

6. *Brundtland Bulletin*, October 1990, p. 11.

7. *Network '92*, April 1991, p. 2.

8. *The Environmental Digest*, No. 68, February 1993, p. 6.

9. Hansen says that this is because he was only in charge of a department, not an agency. But given the fact that his department was dealing precisely with TNCs, we think that he should have been approached.

10. See: P. Chatterjee, 'Who pays for the Earth Summit?', *New Scientist*, London, April 1992, p. 13.

11. Quoted in: T. Harding, D. Kennedy and P. Chatterjee, *Whose Earth Summit is it Anyway?*, Action for Solidarity, Equality, Environment and Development, Amsterdam, June 1992, p. 2.

12. J. Wurst, 'Concern as business helps pay the bill', *Crosscurrents*, CONGO, Geneva, 23–25 March 1992, p. 16.

13. Wurst, op.cit., p. 16.

14. Information from: C.G. Fraser, *Earth Summit Times*, New York, 31 March 1992, p. 6; L. Gould, *Earth Summit Times*, New York, 13 May 1992, p. 17; T. Harding, D. Kennedy and P. Chatterjee, op.cit., p. 1.

8 CHANGING WHAT?

1. S. Schmidheiny, *Changing Course*, Cambridge, Massachusetts, M.I.T. Press, 1992.

2. J.-O. Willums and U. Gölüke, *From Ideas to Action*, Gyldendal, ICC Publishing, 1992.

3. See: W. Schmidt and J. Finnigan, *The Race without a Finishing Line. America's Quest for Total Quality*, San Francisco, Jossey-Bass, 1992.

4. Schmidheiny, op.cit., p. 99.

5. Schmidheiny, op.cit., p. 101.

6. T. Dyllick, 'Question Sensible aux Problèmes Ecologiques', *L'information*, 1990, No. 96, Zürich, Swiss Popular Bank.

7. Schmidheiny, op.cit., pp. 16ss.

8. Schmidheiny, op.cit., p. 83.

9. Schmidheiny, op.cit., p. 7.

10. Schmidheiny, op.cit., p. 72.

11. Schmidheiny, op.cit., p. 34.

12. Schmidheiny, op.cit., p. 1.

13. Schmidheiny, op.cit., pp. 14–15.

14. Schmidheiny, op.cit., p. 74.

15. F. Pearce, 'How Green are Multinationals?', *New Scientist*, London, 23 May 1992, p. 43.

16. Schmidheiny, op.cit., p. 25.

17. B. Kenny, *The Greenpeace Book of Greenwash*, Amsterdam, Greenpeace International, 1992, p. 14.

18. Kenny, op.cit., p. 10.

19. Kenny, op.cit., p. 27.

20. Kenny, op.cit.

21. P. Chatterjee, 'Environmental auditing still awaits its green signal', *Financial Times*, London, 4 February 1993, p. 10.

22. Kenny, op.cit., p. 27.

23. F. Pearce, 'Hit and run in Sarawak', *New Scientist*, London, 12 May 1990.

24. Kenny, op.cit., p. 14.

25. Kenny, op.cit., p. 10.

26. UNCTC, *Benchmark Corporate Environmental Survey*, New York, United Nations, ST/CTC/SER.C/1-4.

27. A. Lees, 'Business wants to help protect the environment but on its own terms', Friends of the Earth UK Press release, 7 May 1992.

28. UNCTC, Benchmark Survey, op.cit.

29. Schmidheiny, op.cit., p. 161.

9 CAN MONEY SAVE THE WORLD?

1. M. McCoy and P. McCully, *The Road from Rio*, Amsterdam, World Information Service on Energy and International Books, 1993, p. 94.

2. McCoy and McCully, op.cit., p. 82.

3. Personal conversation with François Coutu in Rio de Janeiro.

4. Personal conversation with François Coutu in Rio de Janeiro.

5. 'The GEF', *International Environmental Affairs*, 1991, Vol. 3, No. 2, Spring, p. 140.

6. McCoy and McCully, op.cit., p. 82.

7. *World Bank Annual Report*, The World Bank, Washington, DC, 1992.

8. G. Hancock, *Lords of Poverty: the Power, Prestige, and Corruption of the International Aid Business*, New York, Atlantic Monthly Press, 1989.

9. P. Chatterjee, 'How to waste US$5 billion a year', *Guardian*, London, 13 November 1992.

10. C. Duncan, *The World Bank's Greenwash: Touting Environmentalism While Thrashing the Planet*, Greenpeace International, Amsterdam, 1993.

11. Duncan, op.cit.

12. *Tropical Forestry Action Plan*, Rome, Food and Agriculture Organization, 1980.

13. V. Shiva, 'World Bank cannot protect the environment', *Third World Network Briefings*, 1992, No. 8, Penang, Malaysia, p. 23.

14. M. Colchester and L. Lohmann, *The Tropical Forestry Action Plan: What Progress?*, Penang, Malaysia, World Rainforest Movement and *The Ecologist*, 1990.

15. *World Bank Annual Report 1987*, Washington, DC, World Bank, 1987.

16. Hancock, op.cit., p. 158.

17. Shiva, op.cit., p. 22.

18. The British Overseas Aid Programme, *Some Basic Facts*, London, Overseas Development Administration, November 1986.

19. *British Overseas Aid 1985*, Overseas Development Administration, London, 1985.

20. Personal Meeting with USAID at US Embassy, San Salvador, November, 1991. In the 1950s Conrad Black, then President of the World Bank, raised money for the Bank by using the same argument. See: Hancock, op.cit., p. 70.

10 INSTITUTIONAL OUTCOMES

1. See: World Commission on Environment and Development, *Our Common Future*, Oxford, Oxford University Press, 1987, p. 338.

2. *Can We Trust the World Bank?*, Special Issue of *The Ecologist* on the World Bank, Dorset, UK, March/April 1987; B. Rich, 'The "Greening" of the Development Banks: Rhetoric or Reality?', *The Ecologist*, Dorset, UK, March/April 1989, p. 44.

3. *Down to Earth*, June 30, p. 17.

4. M. Khor and C. Yoke Ling, 'The Global Environmental Facility: Democratization and Transparency Principles', *Third World Network Briefing Paper*, 1992, No. 15, Penang, Malaysia, p. 45.

5. Personal communication from Charles Feinstein, operations officer, Global Environment Unit, World Bank, Washington, DC.

6. C. Abugre, 'Why the World Bank Cannot be Relied Upon to Solve Poverty', *Third World Network Briefing Paper*, 1992, No. 14, Penang, Malaysia, p. 43.

7. C. Duncan, *The World Bank's Greenwash: Touting Environmentalism While Thrashing the Planet*, Greenpeace International, Amsterdam, 1993.

8. *Concordare*, 1993, No. 4, p. 1.

9. 'Editorial', *The Ecologist*, March/April 1991, p. 43.

10. *The Network*, July 1993, p. 6.

11. Earth Council press release, San José, Costa Rica.

12. Personal conversation with Josh Karliner, Coordinator of UNCED activities for Greenpeace, September 1992.

13. *Concordare*, 1993, No. 4, pp. 1 and 4.

11   WHAT NOW?

1. The full text is reprinted in: *Third World Resurgence*, 1992, No. 24/25, p. 27.

BIBLIOGRAPHY

Adams, W.M. (1990) *Green Development: Environment and Sustainability in the Third World*, London: Routledge.

Adams, P. and Solomon, L. (1985) *In the Name of Progress. The Underside of Foreign Aid*, Toronto: Energy Probe Research Foundation.

Aggarwal, A. and Narain, S. (1991) *Global Warming in an Unequal World. A Case of Environmental Colonialism*, New Delhi: Centre for Science and Environment.

Amin, S. (1990) *Maldevelopment. Anatomy of a Global Failure*, London: Zed Books.

Bartelmus, P. (1986) *Environment and Development*, London: Allen and Unwin.

Bellow, W. (1982) *Development Debacle*, Washington: Institute for Food and Development Policy.

Brandt, W. (1980) *North–South*, London: Pan Books.

Brown, L.R. (1981) *Building a Sustainable Society*, New York: Norton.

Bungener, P. (1978) *Le Développement In-Sensé. Itinéraires pour un Combat*, Lausanne: L'Age d'Homme.

Chesneaux, J. (1992) 'Bilan de Rio', *Ecologie Politique*, 3–4: 9–26.

Commoner, B. (1971) *The Closing Circle*, New York: Knopf.

de la Court, T. (1990) *Beyond Brundtland. Green Development in the 1990s*, London: Zed Books.

Duncan, C. (1993) *The World Bank's Greenwash: Touting Environmentalism While Thrashing the Planet*, Amsterdam: Greenpeace International.

Friends of the Earth (1977) *Progress as if Survival Mattered*, San Francisco: Friends of the Earth.

George, S. (1977) *How the Other Half Dies*, Penguin: Harmondsworth.

Goldsmith, E. (1988) *The Great U-Turn. De-Industrializing Society*, Hartland: Green Books.

Goldsmith, E., Allen, R., Allaby, M., Davall, J. and Lawrence, S. (1972) *A Blueprint for Survival*, London: Tom Stacey.

Goldsmith, E., Khor, M., Norberg-Hodge, H., Shiva, V. *et al.* (1992) *The Future of Progress. Reflections on Environment and Development*, Bristol: The International Society for Ecology and Culture.

Goodland, R., Daly, H., El Serafy, S. and von Droste, B. (eds) (1991) *Environmentally Sustainable Economic Development: Building on Brundtland*, Paris: UNESCO.

Greenpeace International (1992) *UNCED Undone: Key Issues Agenda 21 Does Not Address*, Amsterdam: Greenpeace.

Hancock, G. (1989) *Lords of Poverty: the Power, Prestige, and Corruption of the International Aid Business*, New York: Atlantic Monthly Press.

Harding, T., Kennedy, D. and Chatterjee, P. (1992) *Whose Earth Summit is it Anyway?*, Amsterdam: Action for Solidarity, Equality, Environment and Development.

IUED (1975) *La Pluralité des Mondes. Théories et Pratiques du Développement*, Paris: P.U.F.

Keating, M. (1993) *The Earth Summit's Agenda for Change. A Plain Language Version of Agenda 21*, Geneva: The Center for Our Common Future.

Kenny, B. (1992) *The Greenpeace Book of Greenwash*, Amsterdam: Greenpeace International.

Kothari, R. (1989) *Rethinking Development. In Search of Humane Alternatives*, New York: New Horizons Press.

Latouche, S. (1991) *La Planète des Naufragés. Essais sur l'Après-Développement*, Paris: La Découverte.

Lerner, S. (ed.) (1991) *Earth Summit. Conversations with Architects of an Ecologically Sustainable Future*, Bolinas, California: Common Knowledge Press.

McCoy, P. and McCully, P. (1993) *The Road from Rio*, Amsterdam: World Information Service on Energy and International Books.

Martinez-Allier, J. (1993) 'Après Rio: L'Ecologisme des Pauvres', *Ecologie Politique*, 6, 43–70.

Nandy, A. (1987) *Traditions, Tyranny and Utopias. Essay in the Politics of Awareness*, New Delhi: Oxford University Press.

Nerfin, M. (1987) 'Neither Prince nor Merchant: Citizen – an Introduction to the Third System, *Development Dialogue*, 1.

Partant, F. (1982) *La Fin du Développement. Naissance d'une Alternative*, Paris: Maspéro.

Polanyi, K. (1944) *The Great Transformation. The Political and Economic Origins of Our Time*, Boston: Beacon Press.

Raghavan, C. (1990) *Recolonization: GATT, the Uruguay Round and the Third World*, London: Zed Books.

Redclift, M. (1987) *Sustainable Development: Exploring the Contradictions*, London: Methuen.

Rist, G., Rahnema, M. and Esteva, G. (1992) *Le Nord Perdu: Repères pour l'Après-Développement*, Lausanne: Editions d'En-Bas.

Rogers, A. (1993) *The Earth Summit. A Planetary Reckoning*, Los Angeles: Global View Press.

Sachs, I. (1992) 'Transition Strategies for the 21st Century', *Nature & Resources*, 28,1: 4–17.

Sachs, W. (ed.) (1991) *The Development Dictionary*, London: Zed Books.

Schmidheiny, S. (1992) *Changing Course*, Cambridge, Massachusetts: M.I.T. Press.

Sitarz, D. (ed.) (1993) *Agenda 21: The Earth Summit Strategy to Save Our Planet*, Boulder, Colorado: Earth Press.

South Center (1993) *Facing the Challenge. Responses to the Report of the South Commission*, London: Zed Books.

South Commission (1990) *The Challenge to the South*, Oxford: Oxford University Press.

Starke, L. (1990) *Signs of Hope. Working Towards our Common Future*, Geneva: Center for Our Common Future.

Timberlake, L. (1985) *Africa in Crisis*, London: Earthscan.

Trainer, D. (1989) *Developed to Death. Rethinking Third World Development*, London: Merlin Press.

UNEP (1981) *In Defense of the Earth. The Basic Texts on Environment*, Founex, Stockholm, Cocoyoc, Nairobi: United Nations Environment Programme.

Wilkinson, R. (1973) *Poverty and Progress*, London: Methuen.

Willums, J.-O. and Gölüke, U. (1992) *From Ideas to Action*, Gyldendal, Norway: ICC Publishing.

World Commission on Environment and Development (1987) *Our Common Future*, Oxford: Oxford University Press.

JOURNALS

The Brundtland Bulletin, The Center for Our Common Future, Geneva, Switzerland.

The Ecologist, Sturminster Newton, Dorset, UK.

The Network (formerly *Network '92*), The Center for Our Common Future, Geneva, Switzerland.

Third World Resurgence, Penang, Malaysia.

INDEX

Acronyms can be found on pages x–xiv.